Physiology Question-Based Learning

This "zentangle" drawing brings to mind a myriad of processes and outcomes which intertwine and complement each other, thus depicting the complex nature of homeostasis

Hwee Ming Cheng
Editor

Physiology Question-Based Learning

Neurophysiology, Gastrointestinal and Endocrine Systems

Springer

Editor
Hwee Ming Cheng
Faculty of Medicine
University of Malaya
Kuala Lumpur, Malaysia

ISBN 978-981-10-0876-4 ISBN 978-981-10-0877-1 (eBook)
DOI 10.1007/978-981-10-0877-1

Library of Congress Control Number: 2016945267

Printed on acid-free paper

This Springer imprint is published by Springer Nature
The registered company is Springer Science+Business Media Singapore Pte Ltd.

Abstract

This Physiology Question-Based Learning (Pq-BL) text focuses on the nervous, hormonal and digestive systems. The previous Pq-BL volume covered cardiovascular, respiratory and renal physiology, and this current Pq-BL completes the selected questions that are asked during the Physiology Quiz International (PQI) competition. The PQI is now in its 14th year and attracts more than 80 medical school teams from over 20 nations.

The illustrations that accompany the questions and explanations in this text include physio-geometric diagrams that summarize essential points in the relevant physiology. The questions are formulated to stimulate students to think and cultivate integrative approaches to understanding physiology. Questions on neuroendocrinology and neural mechanisms in gastrointestinal physiology are included.

The author has taught physiology for 30 years and regularly designs physiology questions to promote conceptual thinking for students of physiology.

Preface

Pleasant words are a honeycomb, sweet to the soul and healing to the bones (Solomon's Proverbs)

This is the second *Physiology Question-Based Learning* (*PqBL*) book and follows on from the first *PqBL* book that contains the physiology of the cardiovascular, respiratory, and renal systems. One theme that was highlighted in the latter was the dynamic aspects of flow, namely, blood flow, airflow, and tubular fluid flow, respectively. This second *PqBL* book covers the other three main physiological systems of neurophysiology, endocrinology, and the gastrointestinal system.

In the former two overlapping areas within neuroendocrinology, we can still think of flow, in this case, of information flow. The human body is "hardwired" by the extensive neural networks that interconnect the central, peripheral, and enteric nervous systems. Neural signals are transmitted as action potentials along myelinated and unmyelinated nerve fibers, and this neural traffic produces motor movements and generates the perception of all sensory modalities.

Information that serves to integrate functions in the multicellular body and maintain homeostasis is also achieved by hormones. We could view these signals as "wireless," circulating in blood as endocrines, in the interstitial fluid as paracrines and autocrines. A large part of these wireless physiologic messages have their controlling server at the hypothalamus in the brain.

Neurophysiology appears to have more eponyms, eg, Monro-Kellie doctrine. To encourage students and teachers to appreciate their physiological heritage, a selected list of eponyms is included as an appendix.

The gastrointestinal (GI) system is a digestive and absorptive biological organ. To perform its diverse functions, from mastication and swallowing to breakdown and absorption of nutrients from carbohydrates, lipids, and fats, both neural and endocrine mechanisms play major roles. The resident GI enteric nervous system (ENS) has been estimated to comprise of just as many neurons within the neural network along the GI tract as there are in the central nervous system (CNS). The CNS and the ENS are functionally linked by the autonomic nervous parasympathetic and sympathetic nerves.

The student might be surprised to be told that the GI could be the largest endocrine organ. The GI endocrine cells are generally not histologically formed as a distinct gland, but are localized in specific segments of the GI tract. Besides major endocrines like gastrin and cholecystokinin that have direct actions related to digestion, paracrines involved in both GI secretory and motility events are also abundant. The GI is also the place where non-adrenergic, non-cholinergic neurons are found.

As in the first book, the last three chapters here contain questions that seek to provide examples of integrative physiology, in neuroendocrinology, and neurogastroenterology. Nerves, hormones, and digestion are also interlinked. For example, glucose in the intestinal lumen releases incretin that stimulates a prior insulin response even before any postprandial hyperglycemia. The parasympathetic nerve acts on pancreatic beta cells during the cephalic phase to release some insulin secretion.

The Solomonic proverb above reminds me of the integration between neurophysiology, endocrine, and GI systems. CNS higher centers involved in speech, language, and emotions are tied to the health of our bones; and we are reminded of calcium homeostasis, intestinal calcium absorption, and the role of osteoblastogenic/clastogenic hormones.

The hand-lettering artwork that begins each of the four parts of this book was drawn by Zhiling, a creative biomedical graduate who also leads our "Thank God It's Thursday" (TGIT) Fellowship evenings in my home.

I hope these *PqBL* pages will help students enjoy physiology and the teachers to continue to be innovative in stimulating physiology learning.

Kuala Lumpur, Malaysia Hwee Ming Cheng

PhysioLego: Concept Building Blocks

Ask students during tutorial with colorful Lego bricks scattered on their shared table:

"What's the next block of information brick you need to stack up and construct the bigger, integrated physiology?"

Using questions, we help to build up, step-by-step, the conceptual knowledge base.

As teachers, we help to construct rationally, with the students, layers of foundational supporting knowledge. Here is an example from cardiovascular physiology (symbolized by red Lego bricks):

- Teach hemodynamics before respiratory physiology with its unique pulmonary vasculature.
- Teach hemodynamics before considering the special features of fetal circulation in utero.
- Build concepts of osmotic pressure before capillary dynamics.
- Build concepts of body fluid spaces (just before) renal physiology.
- Build concepts of blood volume/blood pressure (just before) renal control of blood pressure.
- Build concepts of interrelated sodium balance and volume control (just before) renal regulation of ECF volume.
- Build concepts of second active transport (just before) giving examples of intestinal and renal (re)absorption followed by water osmotic flow.
- Concept of concentration and freely filtered (remaining unfiltered plasma concentration unaltered).
- Concept of parallel blood flow in systemic circulation and ability for local, regional regulation.

- Concept of central brain stem control of arterial blood pressure and local tissue perfusion control (selective, central vasoconstriction with concurrent local vasodilatation).
- Concept of gravitational effects on blood pressure and objective of maintaining arterial blood pressure to adequately perfuse cerebral circulation.
- Concept of arterial elastic recoil and diastolic blood pressure.
- Concept of "upstream" and "downstream" effects of increased/decreased vasoconstriction.
- Concept of venous compliance/capacitance and effects of venoconstriction.
- Concept of central venous pressure/right arterial pressure and perfusion pressure gradient that drives venous return.

We can use the various colors of the Lego bricks to denote the different physiological systems. When asking students to think integratively, building a more wholistic, whole body understanding, we can take a Lego brick (either blue, yellow, green, white, or black, representing respiratory, renal, gastrointestinal, endocrine, and neurophysiology, respectively) and probe accordingly, eg, with a green and a black brick, ask "how is the neural pathway involved in gastric function?" or with a white and black brick, ask "how are the neuroendocrine reflex mechanisms activated in the compensation to restore arterial blood pressure?"

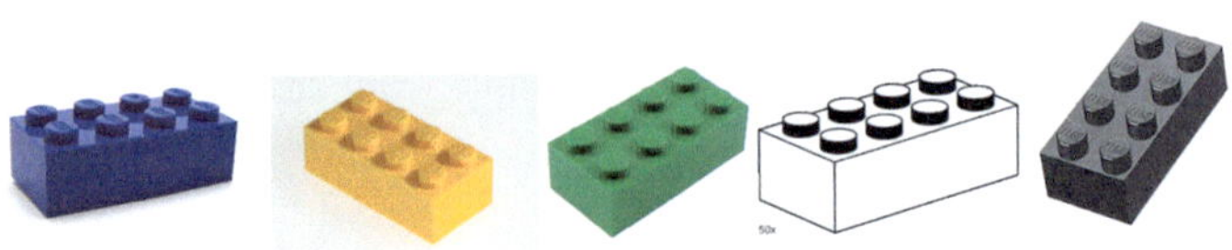

For more Lego concept bricks, "the secretion of the endocrine hormone renin from the kidneys by renal sympathetic nerve during blood pressure control" would include a green renal, a white endocrine, a black neural, and a red cardio Lego brick.

Contents

Part I
Neurophysiology

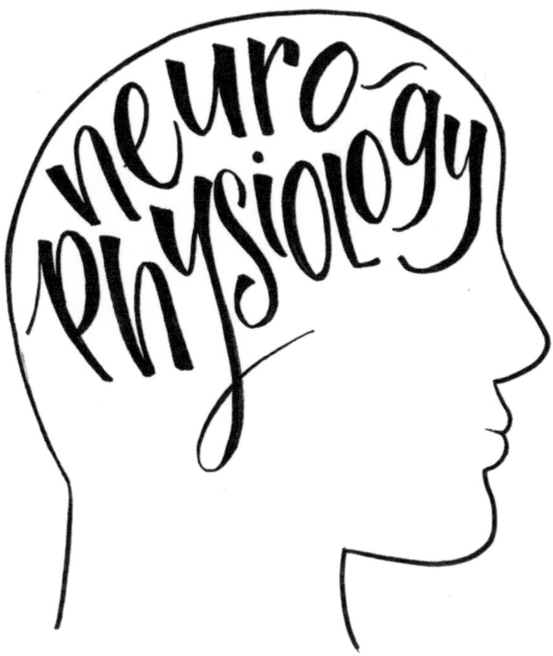

Synaptic plasticity and Religio-physio phenomenon? "…be transformed (*metamorphousthe*) by the renewing of your mind…" Message to the Romans

Introduction: Neural Impulses and Homeostatic Balance

The adjective in the name 'Nervous System' (NS) is certainly not a description of controlled function of neurophysiology in homeostasis. The brain and the spinal cord is termed Central NS (CNS). Projecting from the CNS is the peripheral NS (PNS) which is divided functionally into the somatic and autonomic NS.

All intentional motor movements are initiated in the brain and meaningful execution of specific skeletal muscle groups is by activation of the corresponding alpha motor neuron pool. The autonomic NS (ANS) effects contraction or relaxation of smooth muscles, which are not themselves entirely dependent on extrinsic innervations. Some visceral smooth muscles have intrinsic electrical property that can produce action potentials and contribute to their inherent smooth muscle tone. Unlike the somatic motor action via the alpha motor neurons, which only produces excitation of skeletal muscles, ANS motor smooth muscle events can be inhibitory or stimulatory.

The autonomic sympathetic and parasympathetic nerves also modify the cardiac muscle function. The efferent motor signals that are transmitted along both somatic and autonomic motor fibers are action potentials. The actual ionic events during an action potential and its propagation are predominantly passive sodium and potassium ionic fluxes. The high trans-membrane sodium and potassium concentration gradients of excitable cells are prepared and established by the membrane Na/K ATPase, on alert to stimulus that will depolarize the neuron or muscle cell sufficiently to fire an action potential.

The human body is a sensate, multicellular community. Perception of diverse sensations by mechanoreceptors, chemoreceptors, nociceptors, thermoreceptors and photoreceptors are all conveyed by the same neural signal, an action potential ascending the afferent fibers to the CNS. We see, hear and keep our posture and balance through sensory AP from retinal photoreceptors, hair cell mechanoreceptors in the ear respectively.

The analogy of the neural network to the electrical wiring of a home does not match completely. Neurotransmission of action potentials (AP) are in most cases chemically mediated at the synapses. Both inhibitory and excitatory synapses exist, containing their respective inhibitory and excitatory neurotransmitters. Inhibitory synapses decrease and excitatory synapses increase the likelihood (or excitability) of the post-synaptic neuron to fire AP that will continue the neuronal AP traffic to the target cells. Control of order (and prevention of accidents) is the essential function of road traffic lights. The inhibitory (red) and excitatory (green) signals in the CNS, PNS serve to ensure homeostasis and health.

There is a well described brain-gut axis. This refers to the embedded neural network in the gastrointestinal system, called the enteric NS (ENS). The ENS is largely the final common pathway for the actions of the autonomic parasympathetic and sympathetic nerves on the pattern of motility and secretory activities of the gut.

Specialized neurons also secrete hormones. These neurohormones include the hypothalamic vasopressin and oxytocin and also the spectrum of hypothalamic releasing and inhibiting factors that act on the anterior pituitary.

The discerning student will realize that some aspects of human behavior cannot be easily reduced to neurophysiological and hormonal mechanisms. Human faculties and expressions like emotions (shyness, fear, hate, love), likes/dislikes (Facebook!) memory, learning, understanding, language, music appreciation remains a mystery to be understood in part in future.

Chapter 1
Membrane, Local and Action Potentials

The membrane of all living cells are electrically polarized, with the inner side of the membrane negatively charged and the external surface positively charged. This membrane-associated electrical potential in millivolts (mV) does not conflict with the electro-neutrality of the intracellular fluid and extracellular fluid of the body. Excitable cells like neuron and muscle can be stimulated to produce an action potential. The excitable tissues have a greater polarized cell membrane at rest (resting membrane potential) from minus 60 mV to minus 90 mV.

Action potentials are self-propagating once triggered at sensory neurons, motor neurons and in the muscles. Sequential sodium and potassium ionic fluxes during an action potential depolarizes and then repolarizes the nerve/muscle. Any stimulus that activates an action potential basically causes gated ion channels to open. The local potential change at sensory receptor (receptor potential) then triggers the first action potential in the afferent fibers that innervate the sensory receptors. Action potentials transmitted along somatic or autonomic efferent motor fibers depolarize the nerve terminals to release the respective neurotransmitters.

Remember the membrane and its electrical ability to power the lines of neural communication as the multicellular body responds to changing conditions and maintains homeostasis.

1 State with Reasons, If the Resting Membrane Potential Can Be Determined by Using the Nernst Equation

Answer The Nerst equation calculates for the equilibrium potential in a theoretical condition where only one diffusing ion is considered. Thus the resting membrane potential has to be measured.

H.M. Cheng (ed.), *Physiology Question-Based Learning*,
DOI 10.1007/978-981-10-0877-1_1

Concept The Nernst equation is used to calculate the equilibrium potential (EqPot) for any specific cation or anion. The concentrations of the specific ion in the extracellular and intracellular fluid are used. The EqPot is the calculated electrical membrane potential that will be required to balance or oppose either the influx or the effux of the specific ion.

The direction of the flux of the ion will depend on the direction of the ion concentration gradient across the cell membrane. For example, for sodium it will be a sodium influx if the membrane is easily permeable to sodium and for potassium it will be a potassium efflux. For chloride anion, the diffusion will also be into the cells.

Thus for sodium, the calculated EqPot will be a positive potential to oppose the influx diffusion of sodium cations. For potassium, a negative EqPot will be needed to retard the efflux of potassium cations. Likewise for chloride, a negative EqPot will balance or repel the tendency of the chloride anions to diffuse into and enter cells.

The EqPot values calculated from the Nernst equation for individual ions can be compared with the Measured resting membrane potential (rmp). If the EqPot value for an ion is near the rmp, this gives insight and implies that the cell membrane permeability to that ion is the highest compared to EqPot of other ions that have EqPot values that are further away from the rmp. For example, the rmp for resting neurons is around −70 mV. The EqPot for sodium is positive while the EqPot for potassium is negative and quite close to the rmp value.

This tells us that the membrane of a neuron is most permeable to potassium cations and also infers that the efflux diffusion of potassium down its trans-membrane concentration gradient is the major determinant of the rmp of neurons.

2 State the Quantitative Importance of Direct Action of the Membrane Sodium/Potassium ATPase to the Resting Membrane Potential and to the Action Potential

Answer The electrogenic Na/K ATPase has a small direct effect on the resting membrane potential. During an action potential, the Na/K ATPase is uninvolved since the action potential cycle is due to passive fluxes of ions.

Concept The Na/K ATPase Indirectly contributes to the resting membrane potential. The ATPase exchanger simultaneously pumps sodium outwards and potassium inwards and sustains the trans-membrane sodium and potassium concentration gradients. From the Nernst equation, the EqPot for potassium tells us that the potassium trans-membrane gradient is the major ionic factor that determines the rmp. Quantitatively, the direct electrogenic activity of the Na/K ATPase to the rmp is less than 5 %.

The term ‘action’ potential can mislead the student of Physiology. The ‘action’ does not imply an active, energy requiring event. The depolarization and repolarization phases of an action potential are due to a sudden increase in membrane conductance or permeability to sodium and potassium cations respectively. The trans-membrane sodium and potassium gradients are already available, maintained by the Na/K ATPase membrane pump.

When the cell membrane potential -gated (voltage-gated) sodium cation channels are opened, the sodium ions influx rapidly and depolarize the membrane. The membrane potential changes from a resting negativity to a positive potential. Following on, voltage-sensitive potassium channels open and a reversal of the membrane potential back to a negative value (repolarization) rapidly takes place with a potassium cation efflux.

Thus the Na/K ATPase does not participate in the rapid generation of an action potential. Indirectly, the ATPase sets up the prepared trans-membrane sodium and potassium gradients that is available, 'on duty' waiting for the ionic march influx orders at the commands of an electrical signal, that sequentially opens the sodium and potassium ion channels.

There is a slight hyperpolarization, resulting from an overshoot of the potassium cation efflux during re-polarization phase. After an action potential, the restoration of the normal sodium/potassium trans-membrane concentration gradients is again the primary function of the ATPase.

Note that the number of contributing sodium and potassium cations to a single action potential is very little. What this means is that with One action potential, the sodium and potassium ECF/ICF concentrations do not equalize!

Very soon after a sudden opening of the voltage-gated channels, an almost immediate closing of the channels stop further fluxes of sodium and potassium during depolarization and repolarization respectively.

3 How Does the Conduction Velocity of Nerve Impulses in Autonomic Pre-Ganglionic and Post-Ganglionic Nerve Fibers Compare?

Answer The myelinated pre-ganglionic fibers will have a higher conduction velocity than the unmyelinated post-ganglionic fibers.

Concept Some determinants of conduction velocity of nerve axons include the axon diameter and the myelination of the axons. In unmyelinated fibers, the spread of depolarizing electrical current from active to the neighbouring inactive region is determined mainly by the internal electrical resistance of the axon, which decreases as the axon diameter increases. Thus large, unmyelinated axons have a lower internal electrical resistance and transmit action potentials faster than small-diameter unmyelinated axons.

Myelin is an electrical insulator. Thus the depolarizing current can only affect the membrane potential at the breaks in the insulating layer of myelin (nodes of Ranvier). As a result the action potential is propagated in a discontinuous manner; jumps along the myelinated axon, also called salutatory conduction.

The thickness of myelin and the inter-nodal distance are also directly related to the axon fiber size. Larger axons have thicker myelin and a greater inter-nodal distance. Consequently, large myelinated axons have the highest conduction velocity.

Peripheral nerve fibers are classified according to their conduction velocity into A, B or C fibers, in decreasing conduction velocities. A and B fibers are myelinated and C fiber is unmyelinated. Within the A fiber type are the four sub-categories of A (alpha, beta, gamma and delta) fibers. The pre-ganglionic autonomic motor fiber is a B type and the post-ganglionic motor fiber is an unmyelinated C type.

A different classification of nerves which only applies to sensory, afferent fibers used the notation Type I, II, III and IV, in decreasing conduction velocities. Type IV fibers are unmyelinated (Type C) and transmit sensory impulses from cutaneous thermoreceptors and nociceptors.

The axons with the highest conduction velocity in human Physiology are the motor fibers involved in the control of skeletal muscles (A alpha) and the sensory fibers involved in motor control. The latter afferent fibers are from muscle spindles (Ia) and the Golgi tendon organs (Ib).

4 How Do Changes in ECF Potassium and Calcium Affect the Excitability of Nerve and Skeletal Muscle?

Answer The trans-membrane potassium concentration determines the resting membrane potential and the membrane excitability of nerve/skeletal muscle fiber is related inversely to the stimulus strength required to bring or decrease the resting membrane potential to action potential firing threshold.

ECF calcium affects the diffusion of sodium via its membrane channels during action potential generation.

Concept Changes in ECF potassium levels will change the trans-membrane potassium concentration gradient that determines the resting membrane potential. Hypokalemia increases (hyperpolarizes) the rmp. The membrane excitability is decreased. This explains the occurrence of muscle weakness in hypokalemia.

The situation with hyperkalemia is a little more complex. Hyperkalemia is expected to depolarize the membrane and might then increase the excitability. However, partial depolarization can lead to inactivation of the voltage-gated sodium channels. This then results in a reduced excitability with the same physical symptoms of muscle weakness.

In cardiac muscle, hyperkalemia is a major concern as it can result in asystole via the same suppression of sodium channels.

Calcium in the ECF affects membrane excitability via a different mechanism. A decreased calcium level increases membrane excitability by allowing more sodium influx when the channels are opened. In skeletal muscle a hypocalcemic tetany is the observed consequence (students should not confuse the role of cytosolic calcium in muscle contraction with this effect of ECF calcium).

Hyperventilation has been observed to increase brain excitability in epileptic prone individuals. The respiratory alkalosis results in a pH-dependent hypocalcemia which could explain the increased neuronal sensitivity.

Conversely, hypercalcemia lowers the membrane excitability.

5 What Is the Mechanistic Basis for Describing the Generation of Action Potential as 'All or None'

Answer An action potential is only triggered when the membrane firing threshold is reached. The amplitude of the action potential once activated is always the same.

Concept In excitable cells (nerve and muscle), in general, an action potential is generated when the membrane potential is decreased (less negative) to firing threshold value. At this membrane potential threshold, voltage gated channels are suddenly opened electrically.

Cations rapidly diffuse into the nerve or muscle and produce depolarization. Peak depolarization is reached when the voltage -sensitive channels are acutely closed and inactivated. This rapid sequential opening/closing of the ion channels determine the identical amplitude of every action potential generated.

In vivo, the cationic trans-membrane flux during the rapid depolarization is due to the cationic concentration gradient across the membrane. In most cases, sodium cation is the major depolarizing current. In smooth muscle and at the sino atrial node of cardiac muscle, calcium depolarizing current dominates.

Since the ECF sodium and calcium concentrations are regulated, their respective trans-membrane concentration gradients across the specific excitable cells are also maintained. When the membrane is made permeable to the corresponding cations, sodium or calcium, by the change in membrane potential, the constant ionic concentration gradients also account for the reproducible amplitude of action potentials.

It should be noted that for muscle contraction, both skeletal and cardiac muscles need a spreading depolarizing current in the propagated action potential to effect the mechanical contraction. Increased cytoplasmic calcium in both the skeletal/cardiac muscles are produced by the prior electrical impulses to trigger contraction.

In smooth muscle, pathways for increasing intracellular calcium can occur without the involvement of action potential. These action potential-independent mechanisms involve ligand-gated cell membrane calcium channels and IP3-gated sarcoplasmic reticulum calcium channels.

6 In Vivo, Which Is the Initiation Site for the First Action Potential at the Alpha Motor Neuron?

Answer For alpha motor neuron and all other motor neurons, the axon hillock is the site where the first action potential is triggered and the action potential is then self-propagated down the axon.

Concept The alpha motor neuron represents the final common pathway for the voluntary control of skeletal muscles. There are three directions or sources of synaptic input to the alpha motor neuron. There is the sensory afferent input from the muscle spindles, mechano-stretch receptors also called intrafusal fibers.

There are also spinal interneurons impinging on the motor neuron. One such interneuron is the second intermediary neuron in the bi-synaptic reflex loop that is initiated by afferents from the Golgi tendon organ. The third major modulating input are from descending cortical neurons involved in muscle movement.

The cell body of all motor neurons receives many synaptic inputs, both excitatory and inhibitory. The membrane of the cell body acts like an integrating site that processes the summation of EPSP and IPSP formed on local points on the membrane.

If the net change in the membrane potential is depolarizing, the area of the cell body that has the lowest firing threshold will be triggered to generate the first action potential. Precisely at the axon hillock, the density of voltage-gated sodium channels is the highest. Thus the depolarizing sodium current will activate the action potential first at the hillock. The action potential then self propagates down the axon to its nerve terminal.

An alpha motor neuron and its branches innervates many skeletal muscle fibers, and this represents a functional motor unit. An action potential generated at the axon hillock will lead to electrical impulses transmitted to all the muscle fibers in a motor unit. This means that if there are 70 muscle fibers in a specific motor unit, from one single hillock action potential arising from the cell body, 70 action potentials will be propagated to all the 70 nerve terminals at the neuro-muscular junctions.

7 What Changes in the Membrane Property of Excitable Cells Account for the Absolute Refractory Period of an Action Potential?

Answer The inactivation of voltage-gated sodium channels prevents the generation of a second action potential.

Concept Depolarization phase during an action potential generation occurs when the membrane sodium conductance is suddenly increased as the voltage-gated sodium channels are all greatly opened at the firing membrane potential threshold.

A small amount of sodium cationic influx produces the peak of depolarization. Re-polarization then begins with the concurrent inactivation or closure of the sodium ionic channels and the voltage-gated opening of potassium channels that allows a rush of potassium efflux.

As long as the sodium channels remain shut, a second stimulus will not produce a second depolarization of another action potential. This period of non-responsiveness of the neuron or muscle is termed absolute refractory period. There is a relative refractory period at the tail end of the repolarization phase. This part of the time frame of an action potential coincides with the slight hyper-polarization of the membrane. The sodium channels can at this stage be reactivated to open again to depolarize the cell but however, a greater stimulus than usual is needed since the more negative membrane potential is further away from the firing threshold.

The refractory period of an action potential has at least two physiologic functions. In the cardiac muscle, the ventricular action potential has a unique prolonged depolarization plateau phase (or delayed repolarization). The much longer ventricular myocardial action potential also means that the absolute refractory period is longer. This electrical profile has a protective role in preventing the generation of a second action potential too soon after the first in the cardiac ventricles. As a result, the possibility of a tetanic summation of cardiac muscle contractions is avoided. The normal heart is a rhythmic pump that has to sequentially relax for refilling.

In vivo, action potential traffic is one way (othodromic). How is the retrograde (antidromic) propagation of an action potential prevented in the impulse traffic so that the flow of neural information is focussed and meaningful? When an action potential is transmitted along the axon, the immediate depolarized, active region remains refractory due to the sodium channel inactivation as the next available axonal membrane is being depolarized. Peak depolarization would have occurred further down the axon by the time the voltage-gated sodium channels are reactivated to allow a second depolarizing influx of sodium at the same initial site of the first action potential on the axon.

8 In Vivo, How Are Action Potential Generated in Sensory Afferent Nerve Fibers?

Answer Sensory receptors when stimulated by their respective adequate stimulus elicit receptor potentials which will lead to the generation of the first action potential at the associated afferent sensory fiber.

Concept Our perception of the world and environment is through the diverse spectrum of sensory receptors. These receptors detect different modalities of sensation, from general sensation of touch, pressure, temperature, pain to the special sensation of hearing, balancing, seeing, smell and taste.

Each group of sensory receptors are excited most sensitively to their specific ‘adequate stimulus’. The different types of stimulus that activates the sensory receptors can be grouped under a few categories. One major group is defined by mechanical stimulus which produce receptor potentials in mechanorecpetors. These pressure or stretch-sensitive receptors include the cutaneous touch/pressure receptors, the muscle spindles, the vascular pressure and volume receptors, the hair cells of audition and balance.

Sensory receptors function as transducers and they covert the various stimuli into the receptor potentials. For example, the arterial baro-receptors convert blood pressure stimulus into receptor potentials which then lead to afferent action potentials that are transmitted to the brain stem cardiovascular control neurons.

Chemoreceptors will convert chemical stimuli into the respective receptor potentials. Both the receptors for smell and taste are chemoreceptors. In respiratory control, the carotid and aortic bodies have chemoreceptors.

For vision, light energy is translated into a receptor potential of the photoreceptors. Almost all receptor potentials are depolarizing in nature. The photoreceptors are uniquely hyper-polarized when stimulated by light.

Unlike action potentials, a receptor potential is a local, non-propagated membrane potential change. A receptor potential has no refractory phase like an action potential and can be summated. Some sensory receptors are an integral part of the nerve or naked nerve endings of sensory fibers e.g. nociceptors. For these nociceptors, when the receptor potentials are large enough to depolarize the neighboring membrane to firing threshold, an action potential is then generated and self propagated.

Other receptor potentials depolarize the membrane of the specific receptors sufficiently to release transmitters that then bind to the associated, innervating sensory fibers to elicit an action potential.

Whatever the adequate stimuli, a common signal, an action potential, an electrical message (e-mail), is transmitted to the brain from all sensory receptors via their afferent sensory fibers which enable us to perceive, enjoy and interprete our external surroundings.

9 What Is the Physiological Function of Unstable Membrane Potential in Some Tissues?

Answer Unstable membrane potential produces the rhythmic action potentials that are part of the pacemaker potentials e.g. for cardiac, respiratory and gastrointestinal functions.

Concept Unstable membrane potential means that there is no resting membrane potential. Instead there is a regular, spontaneous depolarization which triggers ac action potential when the firing threshold is reached. This is seen in the cardiac pacemaker potentials at the sino-atrial (SA) node of the right atrium, which sets the resting heart rate. The main ionic mechanism for the non-random, cardiac pacemaker potentials is the concurrent ‘leakage’ fluxes of three ions; sodium (calcium also) influx and efflux of potassium.

Increased frequency of action potentials from the SA node is generated during tachycardia. This is achieved when the rate of spontaneous depolarization is faster (or duration of the pacemaker potential to when the firing threshold is reached is shorter). The pacemaker potential time is reduced during tachycardia by a concurrent increased sympathetic/decreased parasympathetic inputs to the SA node. Sympatheto-mimetic hormones like adrenaline also has the same effect to increase heart rate by speeding up the spontaneous depolarization at the SA node.

Respiratory rate is also due to similar periodic, autonomic discharge of pacemaker-like neurons in the respiratory control centers in the brain stem. Very likely, a group of neurons (pre-Botzinger complex) may function as the respiratory rhythm generators.

When a person voluntarily takes a slow, deep breath, the action potentials are generated in the cerebral cortex and the impulse transmission bypasses the brain stem respiratory neurons enroute to activating spinal motor neurons that control the contraction of the respiratory muscles.

In the digestive tract, there are a number of spontaneous electrical activities of the gastrointestinal smooth muscle. A specific cell, the interstitial cells of Cajal appears to be the source of pacemaker activity in the GI tract. There is the phenomenon of ‘slow wave’ cycle of depolarization/repolarization, the frequency of which varies in different segments of the tract.

During the inter-digestive period or fasting state, there is observed, a periodic (every ~ 90 min interval) spread of peristaltic activity. This spontaneous electrical/mechanical pattern of GI motility is viewed as a ‘house-keeping’ function, to clear undigested, food remnants to prepare the gut for the next intake of food.

10 In Vivo, What Are the Stimuli that Lead to Generation of Action Potentials in Motor Nerve Fibers

Answer Action potentials in efferent motor nerves are triggered by various stimuli from both cerebral neurons and also arising from peripheral sensory afferents.

Concept Students quite commonly think about action potential travelling down its axon from the cell body in an isolated context, one neuron synapsing with another neuron. In the whole intact nervous system, it is essential have a ‘top-down’ view of the way action potentials areenerated.

For voluntary muscle movements, the planning and initiation of a desired action involve cortical association areas, basal ganglia and the cerebellum. The primary motor cortex has neurons that govern the muscles of the body (motor homunculucs). But the initial neuronal activity that causes movement of skeletal muscles do not originate as action potentials in the primary motor cortex.

The activation of action potentials in the primary motor cortex is also influenced by peripheral sensory inputs. From common experience, we make movements in response to our moods and emotions. These changes in human feelings can be due to what we read, hear or see. The recall of memories also obviously affects moods. The profile of the origins of corticospinal tracts provides some perspective on the neural interactions between the motor control and sensory inputs.

Around 30 % of the corticospinal tract neurons arise from the primary motor cortex in the precentral gyrus of the frontal lobe. Another 30 % of the corticospinal tract neurons originate from the pre-motor cortex and the supplementary motor cortex. The other 40 % of corticospinal tract neurons are located in the parietal lobe and primary somatosensory area in the postcentral gyrus.

The Autonomic nervous system (ANS) divisions of the sympathetic and parasympathetic nerve fibers are also motor efferents. The ANS participates in unconscious reflexes that regulate essential functions e.g. blood pressure control. The reflexes are triggered by visceral sensory receptors like the carotid sinus baroreceptors or carotid body chemoreceptors. In the GI tract, parasympathetic vagal motor efferents are stimulated in vago-vagal reflexes when stretch receptors in the digestive tract are activated.

The target tissues of the action potentials transmitted in the ANS are smooth muscles and exocrine/endocrine glands.

Emotions also trigger ANS motor responses effected by a general, dominant sympathetic discharge in what is commonly described as a 'fight or flight response.

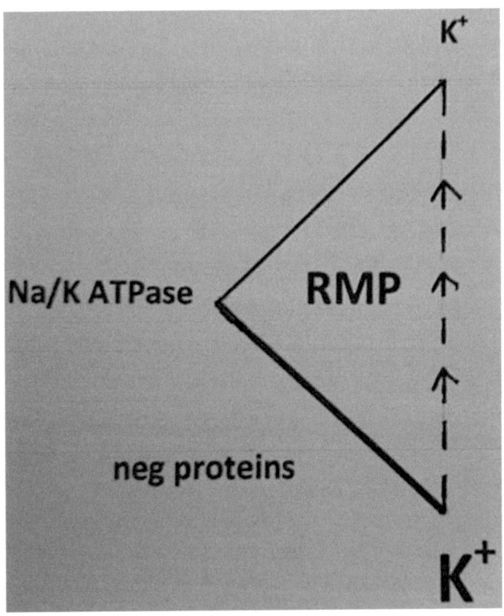

The major determinant of the resting membrane potential (rmp) is the transmembrane potassium concentration gradient. The direct electrogenic activity of the sodium-potassium membrane ATPase contributes just a little to the rmp. The indirect action of Na/K ATPase in setting up the potassium gradient is the main generator of the negative rmp of the polarized cell membrane. The anionic, non-diffusible intracellular proteins have a small role in rmp.

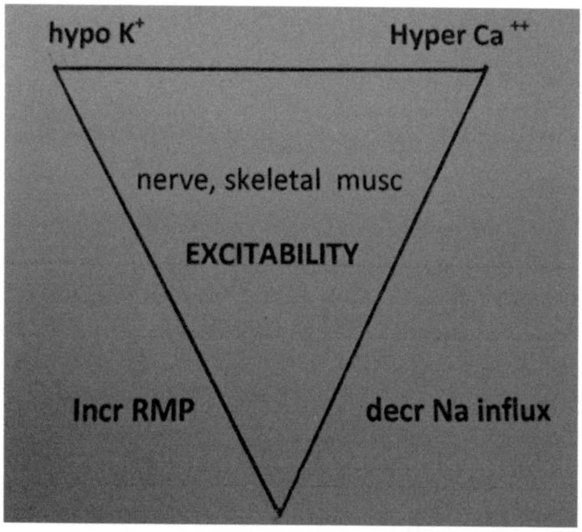

The two extracellular cations, potassium and calcium have low ECF concentrations that are carefully controlled. Changes in the K or Ca concentrations alter the excitability of nerves and muscles. Hypokalemia hyperpolarizes the cell membranes. Hypercalcemia conversely decreases the membrane excitability by interfering with the sodium depolarizing inward current.

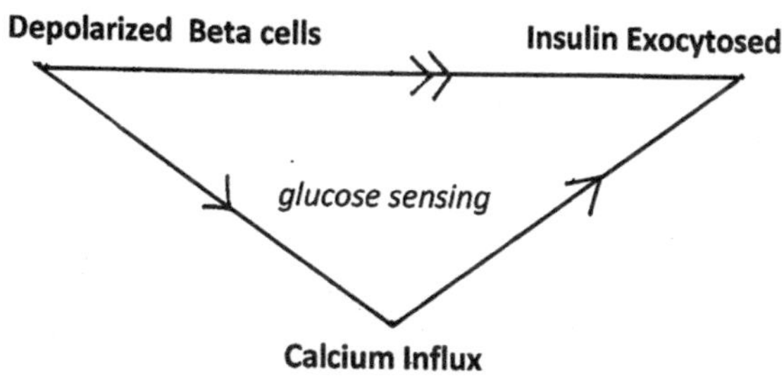

The pancreatic beta cells sense plasma/interstitial fluid glucose concentration. Entry of more glucose by facilitated transport into beta cells leads soon to more cytosolic ATP. ATP-gated potassium channels close and this depolarizes the beta cells. ECF calcium then enters via voltage-gated channels and triggers exocytosis of insulin into the interstitial fluid/capillary blood.

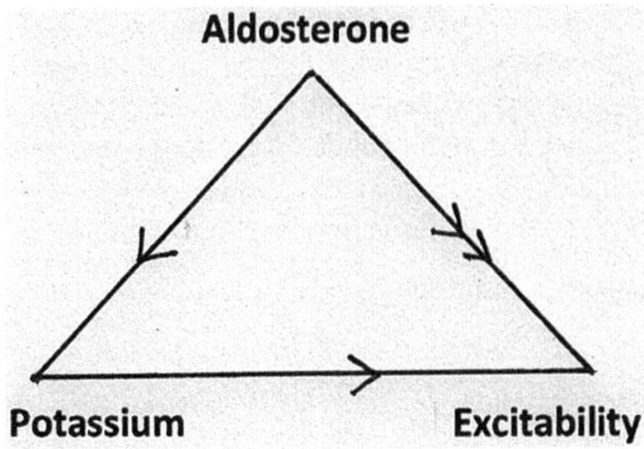

Resting membrane potential of excitable cells are determined by the transmembrane potassium concentration gradient. The adrenal mineralocorticoid, aldosterone prevents hyperkalemia. Hyperkalemia can initially produce abnormal cardiac muscle excitability but if not corrected, the hyperkalemic partial depolarization can cause inactivation of sodium channels and asystole.

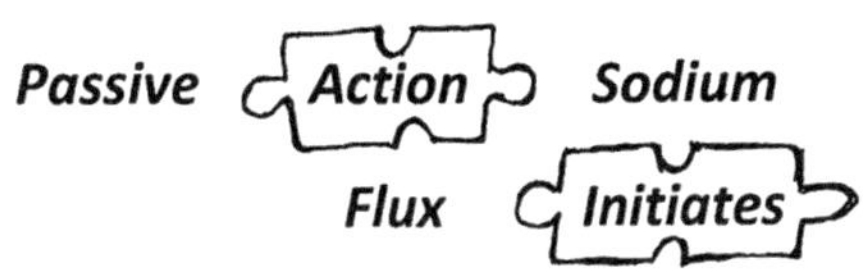

Jigsaw PHYuzzle. Jog and Jig your memory, your thinking. Assemble your physiology. Make a meaningful statement using all these words to define a phase of action potential. Answer (words in reverse) : **Potential Action Initiates Flux Sodium Passive**. The depolarization part of an action potential is due to passive sodium cation influx via voltage-gated channels, down its electro-chemical gradient.

Chapter 2
Synapses and Neurotransmission

Our bodies are programmed for homeostasis. The neural network of the central and peripheral nervous system co-ordinate muscle movement, metabolic, cardio-respiratory and excretory functions. Both afferent signals from sensory receptors and efferent signals to muscles and glands are conveyed as action potentials skipping along myelinated or spreading via unmyelinated nerve fibers. The neuro transmission can be modified along any specific ascending or descending pathways. The points of change or modulation occur chemically at the synapses.

The cell body of the neuron receives multiple inputs, making synaptic contacts with the pre-synaptic nerve terminals. The nature of the neurotransmitter can either increase or decrease the excitability of the cell body to trigger an action potential to be propagated downstream. The quantitative release of the neurotransmitter can also be altered by pre-synaptic effects.

Each cell body of a neuron behaves like a decision making integrator, to increase or decrease the neural traffic of action potentials to the target. The summation of the myriad of synaptic inputs is arithmetic and involve both a higher hyperpolarization or a larger depolarization.

The study of synaptic events allow the specific intervention in dysfunctional neurophysiology to either improve synaptic transmission as in myasthenia gravis or to suppress synaptic signaling e.g. in the management of pain.

1 Does One Action Potential Arriving at a Synapse Always Produce Another Action Potential in the Post-Synaptic Neuron?

Answer One single action potential at the terminal of a pre-synaptic neuron seldom triggers sufficient excitatory neurotransmitters to produce a propagated action potential in the post-synaptic neuron.

H.M. Cheng (ed.), *Physiology Question-Based Learning*,
DOI 10.1007/978-981-10-0877-1_2

Concept The cell bodies of neurons receive numerous inputs from the terminals of many pre-synaptic neurons. At the synaptic points on the cell body, not all the neurotransmitters are excitatory. Chemical synapses are either excitatory or inhibitory. Excitatory neuro-transmission depolarizes locally the membrane of the cell body and brings the membrane potential nearer to the firing threshold. Conversely, at inhibitory synaptic points, the membrane is hyperpolarized and this means that the excitability of that post-synaptic membrane area is reduced.

The post-synaptic potentials are local potentials that can be summated. Thus the net membrane potential change in the overall cell body is the sum of all the excitatory and inhibitory post synaptic potentials that are generated concurrently by the active synaptic points on the cell body. This convergence of many synaptic inputs to one neuronal cell body that then integrates all the incoming excitatory/inhibitory signals is described as a 'many to one' neural architecture.

If the net membrane potential change is depolarizing and the firing threshold is reached, then an action potential is triggered at the axon hillock of the post-synaptic neuron.

There are important exceptions in human Physiology where One action potential in the neuron activates sufficient depolarization to generate another action potential at the membrane across the synaptic junction.

As drawn and pictured in the monosynaptic reflex arc of the muscle spindle stretch response, it appears that one action potential transmitted from the afferent sensory 1a fiber produces an action potential in the cell body of the spinal alpha motor neuron. However since the muscle spindle intra-fusal mechano-receptor fibers are located within a muscle bundle, probably many excitatory post synaptic potentials are simultaneously produced on the cell body of the alpha-motor neuron, when a muscle group is stretched. An action potential is triggered that leads to the reflex muscle contraction.

Two unique central nervous system synapses demonstrate a one pre-synaptic action potential activating an action potential in the post-synaptic neuron. The climbing fiber axon in the cerebellum makes dozens of synapses on a Purkinje cell. In the calyx of Held, along the auditory pathway, the pre-synaptic terminal forms a cap with finger-like protrusions that are in contact with almost 40 % of the post-synaptic neuronal soma. The net summated EPSP generated by the cerebellar or auditory pre-synaptic neuron is large enough to trigger a post-synaptic action potential.

The 'post-synaptic 'membrane at the neuro-muscular junction in skeletal muscle shows end-plate potentials when the excitatory acetylcholine neurotransmitter is released. With a single action potential arriving at the innervating fiber of the alpha motor neuron, sufficient acetylcholine is always released to increase the end-plate potentials to then depolarize the adjacent skeletal muscle membrane to achieve firing threshold.

2 How Can the Excitability of a Neuron Be Progressively Decreased?

Answer By increasing the extent of inhibitory post-synaptic potentials via summation of inputs at inhibitory synapses, the receiving neuron will become less responsive to an incoming action potential at an excitatory synapse located on the same cell membrane.

Concept Both excitatory and inhibitory post-synaptic potentials (EPSP, IPSP respectively) can be summated. Summation of IPSPs makes the membrane potential more negative. This hyper-polarization means a greater subsequent depolarizing current is needed to decrease the membrane potential to firing threshold. The excitability of the neuron is thus reduced.

Summation of IPSP can be effected in two ways. In a single inhibitory pre-synaptic neuron, increasing the frequency of the action potential arriving at the nerve terminal will increase the release of inhibitory neurotransmitter into the synaptic cleft. The local IPSP has no refractory period and can be summated. This frequency -induced effect on a greater IPSP amplitude is termed temporal summation.

Simultaneous discharge of two or more inhibitory neurons can also produce a spatial summation.

An IPSP is produced by either an influx of anions or an increased efflux of cations from the neuron. In vivo, the contributing anion is often a chloride influx and the cation efflux is frequently due to potassium. The nature of the electrolyte is inconsequential. It is the net movement of charges that lead to a greater inner membrane negativity that is responsible for hyperpolarizing the membrane.

IPSP summation, whether temporal or spatial therefore occurs by an increase in the ion movement that results in a more negative membrane potential.

Maintaining normal excitability of neurons in the central nervous system is crucial for normal physiological functions. The uncontrolled burst of neuronal hyperexcitability in epilepsy highlights the essential supplementary role of inhibitory neurons. The purpose of road traffic lights is a good analogy of the co-ordinated task of excitatory and inhibitory neurons. The activity of Green excitatory neurons and Red inhibitory neurons are orchestrated so that the flow of neural information traffic is orderly, directed and focused.

3 How Many Synapses Located Outside the Central Nervous System Are Involved in the Autonomic Nervous Control of Your Heart Rate?

Answer The sino atrial node, the cardiac pacemaker cells have dual innervations by the parasympathetic and sympathetic nerves. Both the autonomic arms have two synapses each outside the CNS.

Concept The first synapse closer to the CNS is called the ganglionic synapse. The sympathetic ganglia is arranged in a chain that hugs the spinal column. The parasympathetic ganglia are nearer the end target organs where the second synapse are located. The second synapses in the autonomic nervous division are sited on smooth muscles, cardiac myocardium or glands.

In both sympathetic and parasympathetic ganglionic synapses, the pre-ganglionic nerve fibers release acetylcholine. The acetylcholine binds to nicotinic cholinoreceptors on the cell body or dendrites of the post-ganglionic neuron.

The ganglionic nicotinic receptors are related but dissimilar to the acetylcholine nicotinic receptors at neuromuscular junctions I skeletal muscle. Pharmacologically, the skeletal muscle contraction is effectively blocked by Curare. The ganglionic cholino-receptors are little affected by curare but Hexamethonium prevents neurotransmission in both the sympathetic and parasympathetic ganglia.

The synapses that connect between post-ganglionic nerve terminal and the end target organs are different from those in the neuromuscular junctions. The points of synaptic contact are along the length of the axon at the cell targets ('en passant' synapses). Synaptic vesicles containing the neurotransmitter are stored in varicosities of the nerve that course along the target tissues.

The major parasympathetic neurotransmitter that acts directly on the target cell is acetylcholine. The cholino-receptor type in the end organs is muscarinic, a G protein coupled receptor ('muscarininc' has nothing to do with muscle but with an agonist, Muscarine).

In the sympathetic post-ganglionic nerve, noradrenaline is the dominant neurotransmitter.

The adrenal medulla also secretes noradrenaline together with the secretion of the major catecholamine, adrenaline which is synthesized from noradrenaline. The sympatha-adreno-medullary axis serves the same physiology roles. Secretion of the adrenal acatecholamines are stimulated by cholinergic sympathetic nerve fibers. This is a mono control unlike in the sino-atrial node, where sympathetic effect increases while parasympathetic action decreases the heart rate.

Another cholinergic sympathetic innervation is the fibers to the sweat glands. There is some report of cholinergic sympathetic synaptic touchpoints on the arterioles in skeletal muscles that could contribute to the pre-exercise vasodilation that prepares the muscle for the increased contractile functions.

4 How Does a Major Neural Mechanism Reduce the Release of Neurotransmitter at Cortical Neurons?

Answer Pre-synaptic inhibition at cortical neurons decreases the release of neurotransmitter at the nerve terminals of central nervous neurons.

Concept Some of the CNS neurons make synaptic connections not at the cell bodies or the dendrites of other neurons. Instead they establish synapses on the pre-synaptic terminal endings. If neuron B and neuron C represent the pre- and post-synaptic neurons respectively, neuron A is the regulatory CNS neuron that impinge on the pre-synaptic neuron B.

If neuron A is an inhibitory neuron, it will hyper-polarize the terminal of neuron B. If an action potential arrives at the pre-synaptic neuron B while neuron A is active, the action potential generated in neuron B will have a reduced amplitude (an exception to the 'all or none' principle of action potential generation?). As a result a lower degree of opening of the voltage-gated calcium channels occurs. Exocytosis of fewer vesicles from pre-synaptic neuron B will lead to a smaller change in the post-synaptic membrane potential on neuron C.

We can imagine that this pre-synaptic inhibition can occur at both excitatory and inhibitory synapses. At excitatory synapses, the excitatory post synaptic potential (EPSP) will be reduced. Similarly inhibition of an inhibitory synapse will decrease the amplitude of the inhibitory post-synaptic potential (IPSP). This slowing down of neural information by pre-synaptic inhibition is like the expected effect of yellow traffic light.

The major inhibitory neurotransmitters in the CNS are the amino acids glycine and GABA (gamma-amino butyric acid). The major CNS excitatory neurotransmitter is also an amino acid, glutamic acid. At the GABAnergic nerve terminals, the enzyme glutamate decarboxylase converts glutamate into GABA.

There are two major subtypes of GABA receptors on the post-synaptic membrane (note in the above scenario of A, B, C neurons, B neuron represents the post-synaptic neuron to the modulating pre-synaptic neuron A). $GABA_A$ receptor binding produces an increase in membrane conductance to chloride anion influx. $GABA_B$ receptor activation hyperpolarizes membrane by promoting an increase in potassium cation efflux from the post-synaptic neuron.

Glycine, like GABA also accounts for direct post-synaptic inhibition. Glycine binding increases membrane permeability to chloride ions. Strychine antagonizes glycine action and the muscular hyperactivity and convulsions in strychnine poisoning highlights the importance of post-synaptic inhibition in normal neurophysiology. Interestingly, in the spinal cord, besides glycinergic, GABAnergic neurons, there is evidence as well for neurons that co-release glycine and GABA from the same vesicles at their nerve terminals.

Surprisingly, the inhibitory neurotransmitter glycine has been found to cross bind and sensitize the receptor response to glutamate, the major excitatory central neurotransmitter!

5 Besides Acetylcholine and Noradrenaline, What Are the Other Potentially Assigned Neurotransmitters?

Answer The other categories of neurotransmitters are mediators in aminonergic, peptidergic and purinergic neurons.

Concept Noradrenaline is a major neurotransmitter under the group biogenic amines. These neuro-active amines include dopamine, adrenaline, noradrenaline (the 3 catecholamines, serotonin and histamine. Serotonin is also called 5-HT (5 hydroxytrptamine, synthesized from tryptophan).

The amino acid neurotransmitters include the major excitatory glutamate and the major inhibitory glycine and GABA.

Some purinergic neurons release ATP together with noradrenaline as found in sympathetic vasoconstrictor neurons.

Nitric oxide represents a unique category of neurotransmitter as it is not preformed and stored in vesicles in the pre-synaptic terminal. It is made on demand, diffuses out and affect not the membrane potentials of the synaptic membrane. Instead nitric oxide modulates the neuronal excitability by effects/affecting on the intracellular biochemical functions in the neurons.

The above groups of neurotransmitters are under the umbrella of small-molecule, rapidly acting transmitters. In contrast, neuropeptides are described as slowly acting transmitters. Neuropeptides are not synthesized at the pre-synaptic terminals and then concentrated in prepared vesicles.

Peptinergic neurons make their neurotransmitters in the cell body using the ribosomal machinery. The vesicles that pack the neuropeptides are then transported slowly by axonal transport to the nerve terminals.

Upon excitation by action potential, the neuropeptide vesicles are exocytosed in like fashion as the small-molecule transmitters. However the vesicles in peptidergic neurons are not recycled but autolyzed. There is also no reuptake mechanism for the peptide neurotrasnmitters after their release into the synaptic space.

A great number of neuropeptides are found both in the gastrointestinal tract and the brain.

The endogenous opiates (endorphin, enkephalin, dynorphin) that modulate pain transmission are neuropeptides.

An important nature of neuropeptide action is that it can produce prolonged actions, lasting for months.

6 How Is Synaptic Strength Enhanced by Repeated Neurotransmission at the Same Synapse?

Answer The synaptic phenomenon that produces this effect is known as long-term potentiation.

Concept Repeated mental activity of a specific kind, e.g. revising over the same neurophysiology for a test, playing frequently the music scales for a violin exam appear to reinforce the synaptic signaling strength across those particular synapses involved in the learning or music playing.

The synapses involved seem to be excitatory synapses that release glutamate. Glutamate post-synaptic receptors are of two mechanistic types; ionotropic and metabotropic. Long term potentiation (LTP) is explained by a cooperative interaction between the two sub- types of ionotropic glutamate receptors namely NMDA and AMPA receptors.

Glutamate released from the pre-synaptc neuron binds to both AMPA and NMDA receptors. The AMPA receptor activation causes a net influx of sodium and depolarizes the post synaptic membrane. The glutamate acts on the NMDA receptor but its effect is only achieved by a co-operative local current from the adjoining glutamate-bound AMPA receptor.

The reason for this is that at resting negative membrane potential, a magnesium ion blocks the NMDA channel and the magnesium is pushed aside only when the membrane is sufficiently depolarized.

This accounts for the association between a high frequency of pre-synaptic action potentials to generate the long-term potentiation change. LTP is considered as a contributing mechanism to learning and memory.

When the NMDA opens, a calcium current influx triggers second messenger pathways that biochemically result in a greater sensitivity of the post-synaptic neuron to subsequent glutamate release.

The cascade of cytosolic reactions in the post-synaptic neuron can also feedback to enhance pre-synaptic glutamate release via retrograde mediators.

Thus repeated synaptic firing along the same set of neurons will facilitate through this LTP, a larger depolarization of the associated post-synaptic neurons.

In pathophysiologic situations of brain cell damage and excessive release of glutamate, excitotoxicity could be the unwanted consequence.

7 How Does the Post-Ganglionic Autonomic Neurons Synapse with Target Cells in the Gastrointestinal Tract?

Answer The efferent fibers of the autonomic parasympathetic and sympathetic nerves pre-dominantly synapse with neurons in the enteric nervous network in the GI tract.

Concept The efferent innervations of target tissues other than skeletal muscle is via the autonomic nervous system (ANS). These tissues include smooth muscle, cardiac muscle and glands. A unique, extensive intra-GI neural network called the enteric nervous system (ENS) is viewed as a subdivision of the efferent autonomic nervous system.

The ENS comprised of two neural networks, the myenteric plexus and the submucosal plexus. Neurons in these plexus synapse with each other or are projected to smooth muscle, glands and epithelial cells.

In general the myenteric plexus controls smooth muscle activity and the submucosal plexus has actions on GI secretory activity. There are neural connections between the myenteric and the submusocal plexuses.

The ENS contains adrenergic and cholinergic neurons as well as neurons that release e.g. neuropeptides, nitric oxide and ATP.

There are two types of reflex arcs in gastrointestinal physiology. The short reflexes are independent of the autonomic innervations and involve only the enteric neurons that bridge the activated sensory receptor (chemoreceptors, osmoreceptors, mechanorecpetors) to the effector cells. The long reflexes from these same receptors are transmitted to the central nervous system by afferent fibers and the effector cells are then stimulated by autonomic nerves acting directly or via the enteric nerve plexuses.

The efferent autonomic neural action is not only reflexly stimulated by various sensations in the gut (gut feelings!). The activity of the brain can have effects on GI activity. Emotional states, hunger, the sight, smell and taste of foods during the cephalic phase of digestion also leads to autonomic stimulation (predominantly parasympathetic) of both motility and secretion in the GI tract.

8 What Factors Limit the Duration of Action of a Neurotrasnmitter in the Synaptic Space?

Answer The three different ways of containing and restricting the actions of neurotransmitters that are released include enzymatic destruction, uptake into presynaptic neurons or adjacent cells and diffusion away from the synapse.

Concept The action of the potent neurotransmitter can be terminated rather rapidly so as to confine their specific action to a particular synapse for an appropriate duration. Among the well established neurotransmitters, acetylcholine alone is inactivated by rapid enzymatic breakdown. The post-synaptic membrane at cholinergic synapses have acetylcholinesterase that acts on acetylcholine to hydrolyze it to choline and acetate. The two products can be recycled into the pre-synaptic neurons to be resynthesized into new acetylcholine.

The synaptic action of monoamines (e.g. noradrenaline) are inactivated by uptake into the pre-synaptic nerve terminals. They may then be re-packed into the synaptic vesicles for subsequent exocytosis and release. Any remnant monoamine that is not taken up by the nerve endings is metabolized by monoamine oxidase or cathchol-O-methyl trasnferase.

Peptide neurotransmitters are not recycled back into the pre-synaptic nerve terminals. The neuropeptides diffuse into the extracellular fluid, away from the synaptic site of receptor action. Further along, the neuropeptides can be digested by extracellular peptidases. The amino acids released can be transported into para-synaptic surrounding cells for reuse in normal metabolic pathways.

Some nerve terminals have a dual neurotransmission, i.e. they co-release two types of neurotransmitters. For example, the parasympathetic nerves that supply the salivary glands co-release both acetylcholine and VIP (vasoactive intestinal peptide). The complementary role of this dual neuro-chemical signals is seen in the action of acetylcholine on the acinar cells to increase saliva flow. Salivary flow is maintained by a concurrent increase in blood flow to the glands. The VIP that is co-transmitted with acetylcholine has a physio-logical vasodilatory action on the vascular arterioles.

9 In the Inverse Stretch Reflex, How Many Types of Membrane Potential Change Are Produced Along the Reflex Arc?

Answer Along the inverse stretch reflex arc, there is one action potential propagated segment and three sites of local potential changes.

Concept The inverse stretch reflex involve the Golgi tendon organs that are arranged in series with the contracting muscle that is attached to the bone at the tendon. The primary function of the Golgi organ is to monitor tension of muscle contraction. The serial location of the Golgi mechano-receptors to the muscles is ideally suited for relaying information of the developing muscle tension.

The Golgi tendon reflex is triggered when the muscle tension becomes extreme and the reflex response is a relaxation of the muscle. The Golgi reflex is protective in such situations besides sending feedback information to the higher centers e.g. cerebellum that is involved in ensuring smooth execution of normal muscle movements.

The receptor potential is generated at the Golgi sensory organ by opening of stretch-gated channels. This leads to a propagated action potential along the axon of the 1b fibers from the Golgi organ. The inverse stretch reflex is a bi-synaptic pathway. An intervening spinal interneuron bridges the afferent 1b nerve endings and the spinal alpha motor neuron that innervates the same muscle group.

The interneuron is an inhibitory neuron and causes an inhibitory post-synaptic potential (IPSP) on the cell body of the alpha motor neuron. The excitability of the hyperpolarized motor neuron is suppressed and a reflex cessation of excitatory action potentials to the muscles results in the relaxation.

The inhibitory interneuron has to be excited by the 1b afferent fibers for it to silence the activity of the alpha motor neuron. Thus an excitatory post synaptic potential (EPSP) is the electrical effect of the afferent nerve on the soma of the interneuron.

Conversely, in the muscle spindle mono-synaptic stretch reflex that results in the muscle contraction, action potential is transmitted along the axon of the alpha motor neuron and finally propagated along the skeletal muscle membrane to the t-tubules for the excitation -contraction coupling. The end-plate potentials generated are normally high enough with just one motor action potential to depolarize neighbouring muscle membrane to firing threshold.

10 Besides Postsynaptic and Pre-Synaptic Inhibition, Are There Other Inhibitory Arrangements in the Nervous System?

Answer Some neurons also demonstrate self feedback inhibition and there is also feed-forward inhibition seen in the cerebellum.

Concept Pre-synaptic inhibition is seen at axo-axonic synapses. Unlike post-synaptic inhibition, where the excitability of the post-synaptic neuron is decreased by hyper-polarization, pre-synaptic inhibition reduces the amplitude of the post-synaptic potential (IPSP or EPSP) through lesser neurotransmitter release without changing the excitability of the post-synaptic membrane.

There is incidentally also pre-synaptic facilitation when the action potential at the pre-synaptic terminals is prolonged and the voltage-gated calcium channels are open for a longer duration.

Pre-synaptic inhibition provides a more precise modulation of neural traffic than post-synaptic inhibition. In the latter, when the excitability of the neuron is changed at the dendritic region or cell body, the neural effects on target cells at all its branch nerve fibers will be equally reduced. Conversely, presynaptic inhibition in a divergent neuron can act to selectively modulate one collateral of the neuron without affecting the other branch fibers.

Descending pathways that modulate pain transmission in the dorsal horn of the spinal cord can act via pre-synaptic inhibition of afferent fibers from nociceptors.

In self-regulatory neuron, e.g. a spinal motor neuron has a recurrent collateral that synapses with an inhibitory interneuron, which then ends on the soma of the same spinal neuron. This inhibitory interneuron is called a Renshaw cell. The inhibitory neurotransmitter is glycine which will self suppress the activity of the motor neuron. Such inhibition via recurrent collaterals is found in the cerebral cortex and limbic system.

Stimulation of basket cells in the cerebellum generates IPSPs in the Purkinje cells. The basket cells and the Purkinje cells are however excited by the same parallel-fiber excitatory neural input. This functional neural architecture, described as 'feed forward inhibition' perhaps serves to contain the level of excitation that is produced by any given afferent neural impulse.

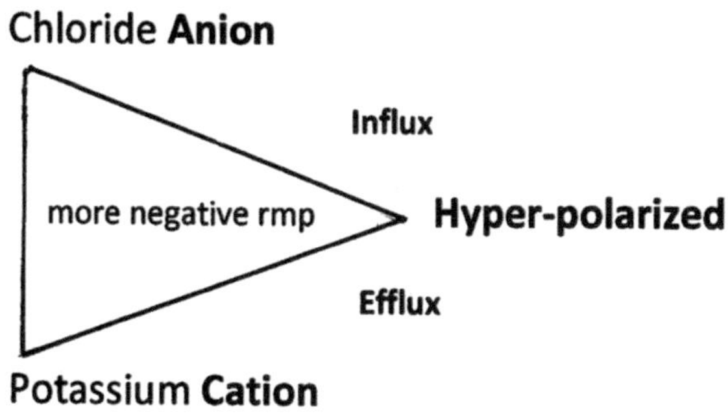

The negative resting membrane potential can be made more negative (hyper-polarized) by either an influx of anions like chloride or an efflux of an intracellular cation like potassium. A hyperpolarized membrane is less excitable as the membrane potential is further away from the firing threshold and hence, a greater stimulus strength is needed to generate an action potential.

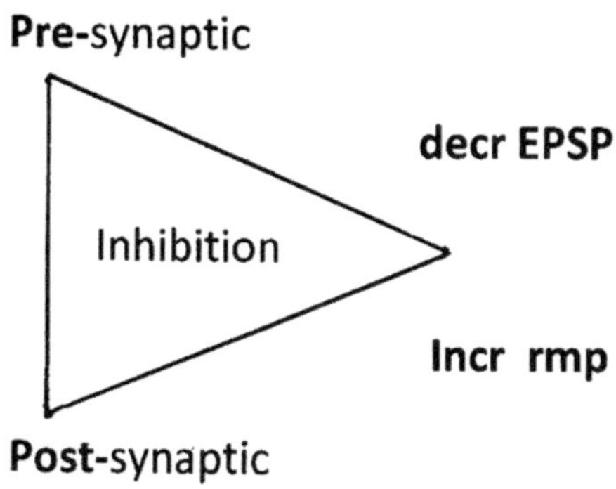

Synaptic inhibition is a major controlling mechanism in the neural communication network. The inhibition can be exerted at the post synaptic membrane of the downstream neuron by increasing (making more negative) the resting membrane potential. The size of the excitatory post synaptic potential (EPSP) can also be reduced by pre-synaptic inputs that decrease the amount of neurotransmitter released.

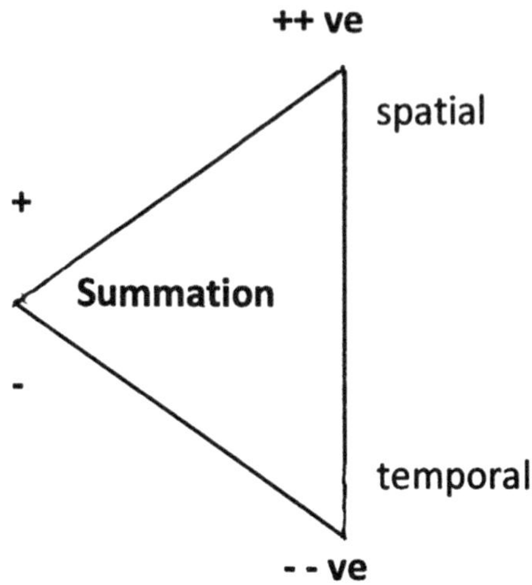

Summation at the cell body is a many-to-one phenomenon, integrating the activities of the myriad of both excitatory and inhibitory synapses. Summation occurs among excitatory or inhibitory synapses. Arithmetic summation of the local post-synaptic potentials also takes place between EPSPs and IPSPs.

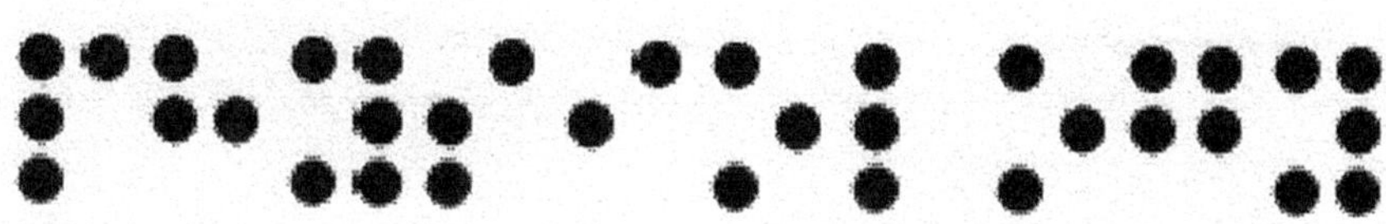

Reading with Braille. The touch sensation, sensory localization and acuity obviously play a role in the transmission of afferent impulses to the brain. The brain normally 'understands' when we hear or see words. It is fascinating that with Braille, the pattern of sensory touch impulses are integrated and interpreted by the language neuronal centers in the cerebral cortex. The Braille above reads PHYSIOLOGY!

Chapter 3
General Sensations and Nociception

Imagine yourself at zero gravity in a pitch black, perfect sound-proof room. It will not be a normal sensory experience. In fact, you will be in a sensory -deficient condition. The comfortable room temperature will still be monitored by your cutaneous thermoreceptors. The usual afferent impulses from touch and pressure mechanoreceptors from both your feet in contact with the ground is absent. The vascular mechano-sensors, baro- and volume receptors are still active to the blood volume and pressure in the circulation.

All sensory impulses are identical electrical, neural messages, action potentials that are sent to the brain to be decoded and interpreted. The body is conscious and able to discern the location or source of the sensory stimulus on its surface. Visceral signals e.g. from proprioceptors and the vascular mechano-receptors are however not perceived.

Pain serves a protective function to alert the person to injury. The leprosy patient with lost of nociception in parts of her body is no longer aware if a harmful insult is present near the insensate region. Nociception also has a psycho-social role. A person who has gone through pain will generally be more sympathetic to a fellow-sufferer. A comforting touch is not merely an activation of the physiologic spectrum of cutaneous mechano-receptors. The touch is perceived by the sensory cortex and the emotional message of the sensation is also understood and appreciated.

1 What Meaningful Information Is Interpreted by Sensory Decoding?

Answer Sensory coding or decoding translates a receptor stimulus to a specific perceived sensation. All sensory systems have four component attributes of their adequate stimuli namely, modality, location, intensity and duration.

H.M. Cheng (ed.), *Physiology Question-Based Learning*,
DOI 10.1007/978-981-10-0877-1_3

Concept In humans, there are four groups of sensory receptors, classified based on their sensitivity to the adequate stimulus, or dominant form of energy that will stimulate a receptor potential in the specific sensory structures. The four receptor types are mechano-, thermo-, chemo- or photoreceptors, the latter responsive to electromagnetic energy.

In 1835, Johannes Muller coined the description 'law of specific nerve energies' to account for the phenomenon that when a particular sensory receptor is stimulated, the sensation or perception elicited is that for which the receptor is histologically designed, regardless of where the stimulus is applied along the nerve.

For localization of a stimulus, the term sensory unit is introduced, which is a single sensory axon and all its branch fibers. The receptive field of a sensory unit is then the space or area from which a precise stimulus will generate an action potential response in the sensory unit. The area innervated by one sensory unit commonly overlaps and interdigitates with the receptive field of other sensory units.

The mechanism of lateral inhibition allows more precise location of sensory stimulus. Impulses from sensory neurons whose receptors are at the periphery of a stimulus are inhibited while action potentials from neurons whose receptive fields are more central to the stimulus are propagated. This contrasting effects enables more precise sensory localization and is the basis for the two-point discrimination cutaneous testing.

The intensity of sensation is also perceived. There are two ways for sensing graduations in the strength of the stimulus. The receptor potential increases with stronger stimulus and this leads to an increased frequency of action potential in a single sensory fiber that is part of a sensory unit. Secondly, as strength of stimulus increases, more receptors are recruited within a sensory axon. In addition, the activation of more than one sensory units by a greater stimulus will also be conveyed and interpreted by the central nervous system as a more intense sensation.

The duration of a stimulus of constant strength results in different responses among the sensory receptors. The frequency of action potential declines with an unchanging stimulus and this is termed sensory adaptation. Receptors are described as rapidly adapting (phasic) or slowly adapting (tonic). The adaptive characteristic is a function of the specific receptors. For example nociceptors are tonic receptors and this 'makes sense' if the receptors are monitoring pain. Muscle spindles, mechano-receptors in muscles, that are involved in posture monitor constant muscle length are also slowly adapting. The Pacinian corpuscles that convey vibratory sensory stimulus are rapidly adapting.

2 What Types of Mechano-Receptors Are Found in the Human Skin?

Answer There are at least four types of pressure and touch receptors in the skin with the eponyms Pacinian, Meisnner, Merkel and Ruffini.

Concept The Pacinian corpuscles and the Meisnner's corpuscles are rapidly adapting receptors. The Merkel disks and the Ruffini endings are slowly-adapting receptors and they respond to sustained pressure and touch.

Meisnner's corpuscles respond to slow vibrations and Pacinian corpuscles are specialized for rapid vibrations and to detect deep pressure. There appear to be no Meisnner's corpuscles in hairy skin, perhaps the hair follicle sensory receptors replacing the transduction role.

The Pacinian encapsulated receptors are surrounded by many layers of flattened fibroblasts. If the associated fibroblasts are removed, by enzyme treatment, the naked nerve endings of the de-sheathed Pacinian corpuscles will be depolarized as long as the mechanical stimulus is applied. In the intact rapidly adapting Pacinian receptor, the fibroblasts enable the Pacinian to respond to rapid vibrations since deformation of the corpuscle depolarizes and reformation immediately after also depolarizes the sensory receptors.

The afferent cutaneous fibers enter the spinal cord via the dorsal roots. The large-diameter afferents then branch on entering the spinal cord and ascend in the dorsal columns to make synaptic connections at the dorsal column nuclei (the cuneat and gracile nuclei) of the medulla oblongata. The second -order neurons exit the dorsal column nuclei as a distinct bundle of nerve fibers named the medial lemniscus. These fibers cross the midline and ascend to the thalamus. From the thalamus, the fibers project to the somatosensory regions of the cerebral cortex.

The sensory inputs from the face are relayed to the brain via the cranial nerve V (trigeminal system). The trigeminal nerve arises at the pontine areas of the brain stem. Near its origin, it expands to form the semilunar ganglion which contains the primary sensory neurons which are analogous to the dorsal root ganglion neurons.

From here three nerves leave the ganglion to innervate the face, namely the opthalmic, maxillary and mandibular. nerves. These sensory nerves transmit information relating to touch, temperature and pain sensations from the face and also from the mucous membranes and the teeth. The facial large A fiber afferents from the mechanoreceptors will transmit information to the thalamus via the medial lemniscus. For thermo- and nociception, the C and A delta fibers from the face will join the spinothalamic tract.

From the thalamus, sensory information from both the large and small afferents will be projected to the face representation area of the primary sensory cortex.

3 What Is the Nature of Cutaneous Thermoreceptors?

Answer Thermoreceptors are present on non-encapsulated free nerve endings of unmyelinated C fibers and myelinated A delta fibers.

Concept In general, cold receptors or cool receptors are on the dendritic endings of A delta fibers whereas warm receptors are on C fibers. There are 5–10 times more cool thermoreceptors than warm receptors.

The actual membrane sensors at the sensory nerve endings are ion channels that belong to a family of proteins called transient receptor potential (TRP) proteins.

Different isoforms of TRP channels have gates that are opened by different temperature ranges.

These channels are non-selective cation channels and when opened, a net sodium depolarizing current occurs. The receptor potential activated then initiates action potential in the sensory afferent neurons which are then transmitted along 'labelled lines' to the specific brain regions where the cool or warm temperature is perceived.

Some of the TRP channels are opened also by chemical ligands. For example TRPV1 (V for vanillod) channels become permeable to cations when bound by capsaicin or ethanol. This dual stimuli response of the TRPV1 thermosensor (activated by temperature >42° C) explains why capsaicin, a component in chilli pepper or ethanol when consumed give a hot sensation.

Similarly, the TRPM8 channel (M for menthol, the ingredient in mint) are activated when the temperature falls below 25° C and this accounts for why menthol applied to the skin feels cool.

Extremes in temperatures are perceived as pain. TRPV2 channels open above a high temperature of 52° C while TRPA1 channels (A for ankrin) are activated below 18° C. These specific TRP channel proteins are found on polymodal nociceptors that respond to mechanical, chemical and temperature extremes. The sensations of freezing cold and burning hot are contributed by these TRP sensor subtypes on nociceptive sensory fibers.

The mechanism of temperature detection probably results not simply from direct physical effects but via a change in the intracellular biochemical reactions in the thermoreceptors. The rate of cytosolic chemical reactions is altered twofold for every 10° C change.

4 Are There Sensory Receptors that Produce the Itch Sensation?

Answer Itch is a specific sensation arising predominantly from the superficial regions of the skin. The receptors are likely bare or naked nerve endings similar to the unmyelinated C-type fibers of nociceptors.

Concept The unmyelinated C type fibers that transmit the pruritic sensation are distinct from those that deliver action potentials from pain receptors. In experiments, skin stripped of the epidermis is very sensitive to pain but is insensitive to itch stimuli.

The reflexes elicited by pruritic stimuli and nociceptive stimuli are also different giving evidence that the C-type fibers pathways propagating the itch or pain sensations are different. Itch provokes the scratch reflex while pain stimuli produce the withdrawal or protective reflexes.

In like manner of 'gating mechanism' for pain transmission in the spinal cord, we know from common experience that scratching relieves itching!. Scratching activates large, fast conducting myelinated afferent fibers that also modulate itch transmission in the dorsal horn neurons at the spinal cord.

Allergic reactions that include itch symptoms are due to histamine and bradykinin. There is evidence that histamine can excite the unmyelinated pruritic sensory C fibers. Bradykinin receptors at bare nerve endings are involved in both pruritogenic and nociceptive responses. Bradykinin receptor antagonist has been shown to reduce the scratching reflex in experimental mouse model.

Causes of itch is not merely restricted to external pruritic stimuli or skin diseases. Itch is also presented in systemic conditions e.g. in chronic renal failure and hepatic problems like obstructive jaundice.

Pruritic afferents travel to the spinal cord in nerve trunks along with other sensory fibers. The second order neurons cross the midline and ascend to the brainstem reticular neurons and thalamus via the ventrolateral spinothalamic tract.

The scratch response is a polysynaptic spinal reflex. In dogs, a limb is positioned at the itch-stimulated area. Then a rhythmic flexion and extension of the limb repeatedly over the pruritic area is aimed at removing the irritant.

5 In the Spinal Cord, How Do Endogenous Opioid Peptides Affect Pain Transmission?

Answer The endogenous opiates can inhibit pain transmission both pre-synaptically and post-synaptically.

Concept Nociceptive sensory fibers end on second-order spinothalamic projection neurons in the spinal cord. Opiate-containing interneurons e.g. enkephalinergic neurons can exert both pre-synaptic and postsynaptic inhibition. Descending serotogenic and noradrenergic neurons can activate these opioidergic interneurons. The afferent nociceptive fibers release glutamate and neuropeptides like substance P that depolarize the spinothalamic projection neurons.

There is evidence that pain perception is transmitted via the thalamus to the secondary somatosensory cortex (SII).

The spino-reticular tract involved during afferent transmission from sites of pain may serve cortical arousal mechanisms and any defense responses of the affected persons.

Pre-synaptically, the endogenous morphine-mimetic neurotransmitters reduce the calcium influx into the nerve terminals. This decreases the exocytosis of the excitatory neurotransmitters at the nociceptive sensory fibers.

Post-synaptically, opioids also hyperpolarize the spinothalamic neurons by increasing the membrane conductance to potassium. This will lead to a depressed excitability of the post-synaptic neuron and a smaller EPSP.

There are also descending fibers that have direct synaptic modulating action on the ascending spinothalamic neurons that convey nociception.

There are also two other sites for the analgesic action of opioids. In the periphery, the inflammatory events result in the release of opioid peptides by immune cells. These opioids can then act to reduce the generation of receptor potentials at the nociceptive receptors.

Morphine injections into the periaqueductal gray matter (PAG) of the midbrain also relieve pain by stimulating descending fibers that inhibit the pain transmission in the spinal cord at the nociceptive nerve terminals in the dorsal horn. These analgesic fibers from the PAG project via brain stem raphe nuclei to spinal neurons.

The analgesic results of acupuncture at the site of pain may work by the same 'gate-control' mechanisms when mechanical or electrical stimulation of large A type fibers reduce synaptic transmission of pain in the spinal cord. Acupuncture point stimulation at a distant site from the pain could perhaps operate by the release of endogenous opioids in the brain.

6 For Visceral Pain, What Are the Characteristics and the Pathways of Transmission?

Answer Visceral pain are poorly localized and the pain transmission is via sympathetic and parasympathetic nerves.

Concept In deep or visceral structures, there is a less dense innervations by nociceptive fibers. There is also a deficiency of A delta nociceptive fibers, so the pain experienced is not rapid, sharp pain. The visceral pain is poorly localized and commonly accompanied by nausea and autonomic symptoms like sweating and blood pressure changes.

Visceral pain is also often referred to other areas. This referred pain is usually to a structure that has developed embryonically from the same dermatomal segment.

The convergence-projection theory of referred pain describes the convergence of somatic and visceral nociceptive fibers on the same second order spinothalamic neurons in the dorsal horn of the spinal cord. For example, pain in the testicles is referred from ureteric distension since the testicle migrated with its nerve supply from the primitive urogenital ridge from which the ureter and the kidney also developed.

There are no proprioceptors in the viscera and also sparse temperature and touch receptors.

Afferent fibers from visceral structures relay to the central nervous system via sympathetic and parasympathetic nerves. Generally the nociceptive fibers from viscera travel in the sympathetic nerves. The exception for those that course in the parasympathetic nerve are nociceptive afferents from the neck of bladder, the prostate, the cervix and the rectum that travel in the pelvic nerves.

The cranial nerves e.g. trigeminal, facial, glossopharyngeal and vagus also have visceral afferents.

The visceral nociceptive nerve endings in the walls of hollow structures are stimulated when these are stretched. Gall stones in the bile ducts and kidney stones in the ureter cause intense pain. Intestinal colic is also produced by the contraction of the dilated portion of the gut before the site of intestinal obstruction.

Ischemic pain is associated with poor coronary perfusion and intense reflex skeletal muscle spasm associated with injuries to bones, tendons and joints. Gastric pain is activated by chemical stimulants in the gastric acid coming into contact with the mucosa of the stomach or the esophagus.

7 Do All Afferent Sensory Impulses Arriving in the Central Nervous System Give Rise to Conscious Sensation?

Answer Not all sensory action potentials are perceived and decoded by the brain as a specific conscious sensation, e.g. baroreceptors that monitor arterial blood pressure.

Concept Another classification of sensory receptors is functionally useful to appreciate conscious and unconscious sensations. This description is based on the Source of the different types of stimuli that are sensed. The teleceptors are receptors that sense events that originate at some distance from the body. Here the retinal photoreceptors, the auditory cochlear hair cell mechano-receptors and the olfactory chemoreceptors serve this teleception.

The exteroceptors as the name implies monitor changes in the immediate external environment. These include pressure, touch and temperature and pain. The four cutaneous mechano-receptors namely Pacinian, Meisnner, Merkel and Ruffini are such physiologic sensors. Cutaneous nociceptors are obviously exteroceptors that inform the body of tissue injury.

The interoceptors sense essential homeostatic parameters like blood pressure/volume (carotid, aortic baroreceptors, atrial volume receptors) and blood respiratory gases (arterial chemoreceptors). The information sensed by the interoceptors are not consciously perceived, and we are unaware of the automatic control of blood flow and breathing.

The mechano-receptors that send feedback information to our brain of the length and tension (muscle spindle, Golgi tendon organ respectively) of our muscles during movement and posture are also interoceptors that ensure effective accurate execution of planned motor activities.

The vestibular hair cells are interoceptors that serve to keep our balance with different postures and during physical activities (dancing to music which our cochlear hair cell teleceptively transduce).

The taste receptors of the tongue are chemo- exteroceptors and we enjoy the pleasure of eating. Interoceptive chemosensors would include the hypothalamic osmoreceptors that detect changes in sodium concentration and the pancreatic glucose sensors at the endocrine islet cells.

The cutaneous cool and warmth receptors are exteroceptors and we adjust our room temperature to make us feel comfortable. The hypothalamic central themoreceptors are however interoceptors and they serve the invisible, unconscious role of maintaining our body core temperature.

Perhaps the one exception to the general principle that interoceptors transmit unconscious sensations are the visceral nociceptive afferents. Pain from our damaged internal organs are obviously experienced and suffered.

8 How Are the Sensory Receptors Involved in Homeostatic Autonomic Reflexes?

Answer The different classes of sensory receptors, mechano-, chemo-, thermo- and photo receptors are all associated with autonomic reflexes that have efferent responses in smooth muscle, cardiac muscle, endocrine/exocrine glands and gastrointestinal enteric neurons.

Concept The different categories of sensory receptors covey a spectrum of stimuli that are perceived by the central nervous system which include touch, pressure, vibration, itch, pain, taste, smell, sight and sound.

Besides these conscious appreciation of meaningful stimuli, there is a variety of unconscious afferent impulses from specific sensory receptors that are part of autonomic reflexes that participate in physiologic homeostasis. A few examples are given here.

In cardiovascular physiology the mechanoreceptors at the arterial wall of carotid sinus and aortic arch and the atria detect fluctuations of blood pressure and blood volume. Afferent impulses are then conveyed to the cardiovascular regulatory neurons in the brain stem. The efferent motor responses are predominantly in the noradrenergic sympathetic nerve fibers that innervate the smooth muscle of arterioles, veins, cardiac muscle and the pacemaker cells in the sino-atrial node. Cholinergic sympathetic activity to the adrenal medulla to affect catecholamine secretion is also part of this autonomic reflex triggered by mechanoreceptors.

In respiratory physiology, breathing rate and depth is controlled by pacemaker-like neurons in the brain stem. The respiratory neurons in turn receive afferent signals from arterial chemoreceptors (aortic and carotid bodies) that serve to maintain a constant partial pressure of carbon dioxide and oxygen in arterial blood. Hypercapnia, acidosis or hypoxemia are the three adequate stimuli for generating action potentials in the afferent fibers from the chemoreceptors.

The 'automatic' reflex via the respiratory neurons will be an increase in alveolar ventilation, with a greater tidal volume and frequency of breathing. This compensatory motor response in this case does not use sympathetic efferent nerve as the inspiratory muscles are skeletal and spinal alpha motor neurons innervate the diaphragm and the external intercostal muscles.

Both our pupils reflexly constrict and decrease in size when light is shone directly into one eye. This photoreceptor activated direct and consensual papillary reflexes are due to the post-ganglionic parasympathetic innervation of the circular smooth muscles of the iris. A collateral of the optic nerve terminates in the oculomotor nucleus from where the pre-gangionic parasympathetic fiber travels to the ciliary ganglion.

The iris has a dual autonomic innervations and reflex dilation of the pupils are effected by sympathetic activity to the radial dilator muscles.

Hypothalamic central thermoreceptors respond to increased core body temperature by reflexly stimulating sweating and cutaneous vasodilation to lose heat. The efferent fiber activities that cause these effects are respectively, increased sympathetic cholinergic to sweat glands and a decreased noradrenergic vasoconstrictor sympathetic to the skin blood vessels.

9 In Vascular Smooth Muscle, How Do Sensory Receptors Serve Their Physiologic Role in Regulating Regional Blood Flow?

Answer The smooth muscles of the main resistance vascular segment, the arterioles are sensitive to local mechano-, chemo-, and thermo-stimuli and adjust their vascular radius accordingly.

Concept In certain organs, blood flow to the tissues is autoregulated relatively constant when there are fluctuations in the arterial perfusion pressure. This intrinsic autoregulation of blood flow is a distinct property at the cerebral. Coronary and the renal circulations. The myogenic mechanism involving mechanoreceptors art the arteriolar smooth muscles mediate the regulation.

Should the perfusion pressure increase, stretch-gated calcium channels are opened at the arterioles. This produces an arteriolar contraction that increases the vascular resistance. The blood flow rate is maintained.

During physical activity, blood flow to the cardiac and skeletal muscles are increased to meet the greater metabolic demands. The coronary and skeletal arterioles both have chemoreceptors that respond to the changing local factors in the tissues. Providentially, these local metabolic factors are vasodilators. These include local tissue hypoxia, hypercapnia and acidosis. Increase in tissue temperature also enhance the vasodilation. This physiologic phenomenon is called metabolic hyperemia.

In the unique case of pulmonary blood vessels, the sensors that respond to hypoxia does not result in vasodilation. A peculiar pulmonary hypoxic vasoconstriction (hpv) is the observed effect. This hpv obviously does not need to serve a compensatory metabolic function in the lungs but is directed towards maintaining optimal ventilation/perfusion matching at the pulmonary alveoli. The hpv is due to a hypoxia-induced reduction in the membrane conductance to potassium. The pulmonary vascular smooth muscles are depolarized and voltage-gated channels open to allow a calcium influx that activates the hpv.

During injury, tissue trauma involving blood vessels, hemostasis is initiated by an immediate vascular constriction. This smooth muscle response could be triggered by both mechano- and chemosensors during the tissue damage.

The circulation is a closed system. There is a need to selectively re-distribute the fixed blood volume to priority organs when e.g. during hypovolemia of body fluid loss. Smooth muscles participate in both the generation of afferent impulses and the compensatory efferent response during the circulatory adjustment. Mechanoreceptors at the arterial baroreceptors, the volume sensors at the atria, pulmonary vessels and intra-renal baroreceptors provide the sensory input to trigger either a selective sympathetic vasoconstrictor action to non-essential organs or the release of renin respectively. The former autonomic neural reflex acts on vascular smooth muscles at the splanchnic, cutaneous and renal circulations. The arteriolar smooth muscles in the brain and the heart are spared or unaffected to ensure adequate cerebral and coronary perfusion during the hypovolemia.

10 Are Sensory Receptors Involved in Somatic Reflex to Skeletal Muscles?

Answer Sensory receptor activation do lead to contraction and also in some situations, relaxation of skeletal muscles.

Concept The somatic peripheral nerve division consists of a single alpha motor neuron between the central nervous system and the skeletal muscle. Excitation of the motor neuron always leads to muscle contraction.

The muscle proprioceptors, muscle spindles and the Golgi tendon organs are mechanoreceptors that monitor static, dynamic muscle length changes and muscle tension during contraction respectively. Both these proprioceptors send afferent impulses to the spinal cord to excite or inhibit alpha motor neurons. Together their differing reflex effects serve to maintain posture.

Skeletal muscle tone is defined as the resistance to passive stretch that the reflex muscle contraction produces.

Nociceptive fibers activated by an acute injury and pain triggers a protective flexion, withdrawal reflex. Afferent volleys excite motor neurons to flexor muscles and simultaneously, interneurons that inhibit motor neurons that supply the extensor muscles are inhibited. This concurrent different synaptic effects at flexor and extensor alpha motor neurons during a reflex generated by stimulation of nociceptors is termed reciprocal inhibition.

Chemoreceptors, both peripheral and central that sense blood carbon dioxide levels are reflexly involved in the regulation of respiration, via efferent impulses to the skeletal muscles of inspiration transmitted from the brain stem.

During exercise, there is the fascinating observation that the hyperventilation during the physical activity is associated with relatively unaltered partial pressures of CO_2 and oxygen in the blood. Blood pH is also unchanged. One proposal is that chemoreceptors in skeletal muscles sense the local tissue metabolic changes and this activates afferent excitatory signals to the respiratory neurons in the brain stem.

Chemoreceptors give us perception of taste and deliciousness to enrich our eating. The muscles of chewing are skeletal. Sudden oral taste of bitterness can reflexly makes us spit out the undesirable food.

The swallowing neuronal center is located in the brain stem medulla. The natural swallowing reflex is initiated by mechanoreceptors at the back of our mouth, near the pharynx. Afferent sensory impulses are conveyed to the swallowing center via the glossopharyngeal and vagus nerves. Efferent impulses that coordinate swallowing are sent to the striated muscles of the pharynx and the upper third of the esophagus.

Involuntary shivering of skeletal muscles generate heat as a compensatory response to cold exposure. Hypothalamic l thermoreceptors play a central role is this thermoregulatory reflex.

The striated muscles of Smiling are obviously linked to emotional stimuli. Our moods are affected by what we see, read, hear or what we remember. Only Humans Smile.

Neural traffic and Traffic Lights. Red symbolizes the Inhibitory synapses. Green symbolizes the Excitatory synapses. As red is above green, this also highlights that there are perhaps just as many or more Inhibitory synapses than Excitatory Synapses involved in the homeostatic function of the nervous system. Yellow symbolizes the integration of the numerous Inhibitory/Excitatory inputs by the entire post-synaptic membrane of the neuron cell body.

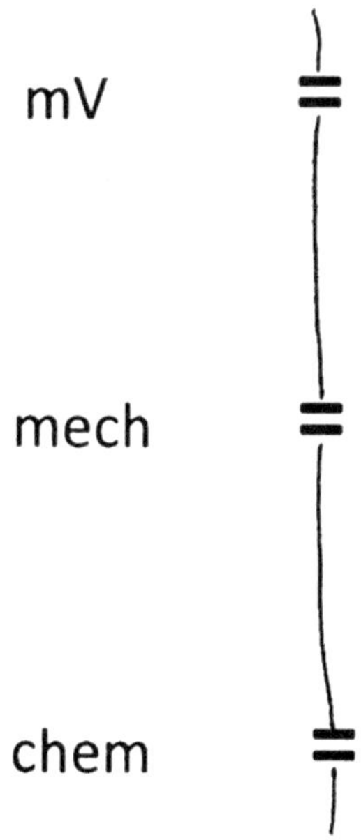

Each sensory receptor cell is like a gated community, an elite, electrolyte gated community. Fluxes of ions move across the membrane via ion channels which are often gated. The gating is controlled either by changes in membrane potential (mV), mechanical stretch or by chemicals/ligands. For photoreceptors, light activates closing of sodium channels and this uniquely hyperpolarizes the photo receptors.

>> What?

Sensation >> Where?

>> wwW

>> w**w**w

Several aspects of sensory decoding. Different adequate stimuli are all transduced into the universal neural message as action potentials (AP). The brain decodes or interpretes the received AP as meaningful information about the sensory modality (what), locality (where), intensity (wW) and acuity (w**w**w).

Fast pain >>> **myelinated >>> *A delta***

Slow pain > > > un**myelinated > > *C***

Sharp acute pain is transmitted by A myelinated fibers while dull, unfocussed pain has a slower transmission along smaller, unmyelinated sensory fibers.

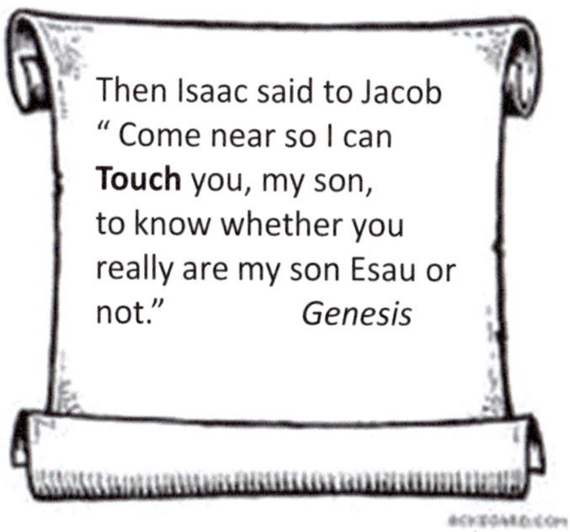

The sensation of Touch is mediated by a spectrum of cutaneous receptors. Touch is an essential factor that enhances human relationships.

calcium 'gate-crashes'
(vCa channel and Ca gradient)
at the terminals and triggers
a burst of nervous activities,
some positive (EPSP),
some negative (IPSP)!

Influx of calcium ions when the nerve terminals are depolarized triggers release of excitatory or inhibitory neurotransmitters, which then bind to the membrane of the post-synaptic neuron.

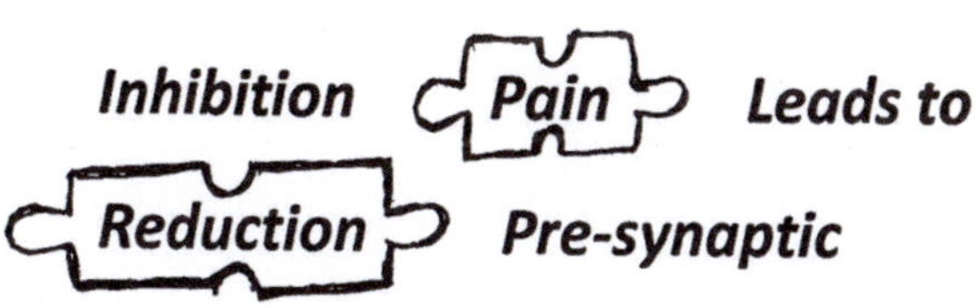

Another PHYuzzle. Answer (words in reverse): **Reduction Pain Leads to Inhibition Pre-synaptic**. The modulation at the spinal cord, of nociceptive afferent impulses transmitting to conscious perception of pain intensity is called 'gate-theory' of pain. Descending inhibitory fibers that secrete opioid neurotransmitter produce pre-synaptic inhibition to lessen pain.

Chapter 4
Special Sensations

The sensory neurophysiology of our perception of sight, hearing, balance, taste, smell are called special senses. In reality, all the general sensory receptor functions (touch, pressure, temperature) are special in their own way. This ‘special’ definition applies also to the circulatory system where every regional vasculature (cerebral, coronary, pulmonary, skeletal, splanchnic, cutaneous) have their unique, local regulatory mechanisms for blood perfusion.

The retinal photo receptors, hair cell receptors of hearing and equilibrium in the ear, olfactory and gustatory receptors are specially dedicated to light, mechanical and chemical stimuli respectively. Life is special and pleasant when we can see, enjoy good music, walk upright, smell a hibiscus and bite into a tasty sushi roll.

The sensory receptor potentials of these structures are special with regards to the ionic basis of their generation. The cone and the rod photoreceptors are hyperpolarized by light. The hair cells from the cochlear and vestibular apparatus are depolarized by a unique potassium influx. The taste sensations involve several transduction mechanisms that produce the perception of bitter, sour, salty, sweet and umami (and recently, fatty taste of delicious roast pork).

The olfactory receptors are primary afferent neurons, making these specialized neurons, the only points of contact of the nervous system with the external environment. Boy-girl relationships are presumably enhanced by those who believe in perfume.

1 How Is the Spectrum of Taste Sensations Transduced?

Answer Depolarization of the taste chemoreceptors involve both inotropic (salty, sour) and metabotropic (sweet, bitter, umami) receptors.

Concept The sense of taste (gestation) in provided by the presence of about 10,000 taste buds on the tongue. Each taste bud is innervated by approximately 50 nerve

H.M. Cheng (ed.), *Physiology Question-Based Learning*,
DOI 10.1007/978-981-10-0877-1_4

fibers. Each nerve fiber receives input from an average of five taste buds. There are between 50 and 100 taste cell receptors that respond to taste stimuli or tastants. There are epithelial basal cells surrounding the taste bud. These basal stem cells differentiate into new taste cells and replace the old gustatory cells with a half life of about 10 days.

Taste cells extend from the base of the taste buds to the taste pore, where microvilli are exposed to tastants dissolved in saliva and mucus.

Salty taste, triggered by sodium chloride is mediated by amiloride-sensitive epithelial sodium channels called ENaC. Influx of sodium depolarizes the salt receptor membrane to generate a gustatory receptor potential.

The sensation of sourness is activated by protons. The hydrogen cations appear to block a potassium sensitive channel. This reduces the membrane permeability to potassium efflux and the sour taste receptors are depolarized. Depolarization of both salty and sour taste sensors will open voltage-gated calcium channels. The acute calcium influx leads to exocytosis of neuro-transmitters in the sensory receptors onto their afferent innervating fibers.

For perception of sweet, bitter and umami taste, metabotropic receptors are activated which are coupled to a G protein, gustducin. Gustducin increases the level of inositol triphosphate (IP3) which then causes an increase in cytosolic calcium, to trigger the release of neurotransmitters from the taste receptors. G protein-linked signaling can also act via cytosolic cyclic AMP changes to decrease potassium conductance and depolarize the taste receptors. The umami taste is effected by a glutamate metabotropic receptor.

The gustatory receptors in the tongue are innervated by branches of the facial and glossopharngeal nerves. The chorda tympani branch of the facial nerve (VII) supplies the anterior two-thirds of the tongue and the lingual branch of the glossopharygeal nerve (IX) receives afferent input from the posterior one-third of the tongue.

The sensory fibers from taste-responsive areas other than the tongue travel in the vagus nerve to the brainstem. These three sensory fibers of taste project to the gustatory area of nucleus of the tractus solitarius (NTS). The second order neurons in the taste pathway leave the brain stem NTS and synapse in the thalamus. Third order neurons from the thalamus then terminate at the gustatory center in the ipsilateral cerebral cortex (anterior insula, frontal operculum).

The students should note that taste afferents from the NTS play a role in the visceral reflexes during the cephalic stage of gastrointestinal secretions.

2 How Is the Wide Variety of Different Smells Discriminated by the Brain?

Answer Odor discrimination is coded by the pattern of neural activity in the olfactory bulb glomeruli.

Concept Smell also affects the pleasurable sensations of eating as experienced by a reduction in taste when we have a cold and our nasal smell receptors are less sensitive. The temperature and texture of foods also influence the perception of taste e.g. enjoying hot soup and crispy curry puffs respectively.

The olfactory mucosa has three types of cells, namely basal cells, supporting cells and the olfactory sensory receptors. Uniquely, the basal cells are precursor stem cells for the olfactory receptors which are replaced every 60 days.

An olfactory sensor is actually an afferent neuron. The axons that extend from the olfactory neuronal cells are gathered together to form the olfactory nerve (1st cranial nerve). The olfactory neuron is thus the only part of the nervous system that is exposed to the external environment!

The receptor sensing portion of the olfactory cell is an enlarged knob with cilia extending into the nasal mucosa. These cilia contain the sites for binding of odorants.

The human nose has in total about 5 million olfactory receptors and there are estimated a thousand receptor types that sense different odorants.

Olfactory receptors are G protein coupled. The cascade of cyclic AMP-dependent cytosolic reactions are activated. Then non-selective, cAMP-gated ion channels are opened. This leads to a net cation influx which generates the receptor potential in the olfactory sensory neuron.

The afferent fibers synapse in the olfactory bulb, which has a complex interconnecting functional layers of neurons, similar to the architecture of the retinal neuronal layers.

Each olfactory bulb is lined by 'little balls' (glomeruli) of neural junctions.

The afferent fibers synapse with the mitral cells in the glomeruli. Each glomerulus receives afferent sensory impulses from only one odorant component stimulus. Various parts of an odor are thus detected by different olfactory receptors and sorted out at the glomeruli into separate 'smell files'. The olfactory glomeruli, as the first smell information relay stations to the brain already serve to discriminate odor. The mitral cells with their glomeruli project to different points in the olfactory cortex. The cortex can distinguish at least 10,000 different scents via this glomeruli sorting.

From the olfactory bulb, nerve fibers travel in two separate pathways. One route is to the hypothalamus and amygdala which is likely associated with smell and behavioral responses. The other olfactory neural traffic ends in the cerebral cortex via the thalamus where fine discernment and perception of smell is interpreted.

In mammals, though not as distinct in humans, the nose also contains the vomeronasal organ (VNO) which detects pheronomes. VNO sensory neurons project to the olfactory bulb and fibers from there synapse with neurons in the amygdala and hypothalamus. The appeal of advertisements for perfumes is obviously a testimony to the evidence that there are pheromone-like stimuli in humans that contribute to the association between smell and sexual behaviour.

3 How Do Sensory Receptors in Our Ears Serve to Give Us the Sense of Balance and Contribute to Maintenance of Posture?

Answer Vestibular information is integrated with sensory information from the eyes, cutaneous receptors, propioreceptors in joints and muscles to the vestibular nuclei in the brainstem and the cerebellum for maintaining balance and a desired posture.

Concept The vestibular apparatus of the inner ear is essential for equilibrium and for monitoring the position and motion of the head. The vestibular sensory apparatus consists of two functional structures near the auditory cochlear – the semicircular canals and the otolith organs.

As for the cochlear, all components of the vesticular apparatus contain endolymph and are surrounded by perilymph. The sensors in the vestibular apparatus are mechanoreceptors, hair cells that are bent by movements of the endolymph fluid. Receptor potentials can be depolarizing or hyperpolarizing depending on the direction of deformation of the hair cells. Most of the information from the vestibular system, relating to maintenance of posture and balance does not reach consciousness.

The 3 semicircular canals detect rotational or angular acceleration or decceleration. We are conscious of being spun. The otolith ('ear stones') organs serve to provide information of the position of the head relative to gravity. The 2 otolith sensors, the utricle and saccule also detect linear acceleration and decceleration. The hairs (sterocilia and kinocilium) of the hair receptor cells are embedded in an overlying gelationous layer that contains the calcium carbonate otoliths. Movement of the crystalline layer triggers receptor potentials in the saccule and the utricle.

The utricle detects changes in head position displaced away from the vertical and horizontal acceleration/deceleration. The saccule senses instead movement in head position away from the horizontal and thus when enjoying bouncing on a pongo stick or being transported in an elevator, the saccular hair cells are activated.

Posture is maintained as you rise up to walk. The crystal-heavy gelatinous layer of the utricular cells, because of its greater inertia, lags behind the endolymph as you start to walk forward. The hair bends backwards, opposite to the movement of your head. Action potentials are generated. When walking at a constant pace, the utricle cells ceases to send sensory impulses. When you stop walking to rest, again the otolith sheet continues to bend the hairs forward when the head is already stationary.

Tilting the head in any position from the vertical e.g. during a warm up exercise of bending body backwards and then touching one's toes stimulates the utricle hair cells. Our posture is maintained throughout the physical activity as reflexes involving vestibulo-spinal pathways to anti-gravity muscles ensure our balance.

Sensory information from the muscle/joint proprioceptors and receptors of the skin in contact with the floor are also integrated with vestibular signals in the brain stem and cerebellum to co-ordinate the appropriate orientation and motion of the body in space and time respectively.

4 How Do Receptors in Our Ear Function to Give Us a Stable Vision During Walking or Jogging?

Answer The vestibular-ocular reflexes that control our eye movements via the extraocular muscles stabilize the visual field on the retina during physical activity that moves our head.

Concept Signals from the semicircular canals regulate eye movements. The sensory hair cell mechano- receptors of the semicircular canals are located at the ampulla of the three canals, arranged at right angles to each other.

The mechanism for the generation of receptors in the vestibular system (semicircular canals, saccule, utricle) are similar to that for the auditory hair cells of the cochlea. Uniquely, the endolymph that bathes the hair cells is potassium rich. Thus a potassium influx depolarizes the hair cells.

Turning the head to the left horizontally will excite hair cells and the innervating afferent fibers of the vestibular branch component of the cranial VIII nerve. Simultaneously, the hair cells in the corresponding canal in the right ear will be inhibited. Thus the sensory outputs from the two ampullas act as a couplet to give the perception of the angular or rotational head movement.

Head movements in other planes will likewise differentially activate and suppress hair cells in each semicircular 'canal couple'.

When the head is moved to the left, a vesticulo-ocular reflex results in the eye turning to the right. This reflex is to permit the gaze to remain steady so that the retinal image is stable.

Interestingly, if the vestibular system is damaged bilaterally, the person might be unaware of any sensory dysfunction. However, if the visual cues are blocked by blindfolding, the person has difficulty in maintaining her posture and will not be able to walk across a compliant surface like a mattress without falling over!.

Motion sickness is not a reflection of any defect in the vestibular system. Rather it is caused by a dissynchrony between sensory impulses from the vestibular apparatus and that provided by other afferent inputs from the retina and the position sense proprioreceptors.

The vestibulo-ocular reflexes can be assessed in the laboratory by using the phenomenon of nystagmus. The Barany chair is used to rotate a patient. Rotation to the right produces a normal right rotatory nystagmus. When the Barany chair is stopped, a left postrotatory nystagmus is observed.

A caloric thermal stimulation can also be used to test vestibulo-ocular reflex. Each ear is rinsed separately with warm water. This will stimulate endolymph flow on the side of the ear inundated with warm water. A nystagmus is produced in the treated ear if the reflex is intact.

5 What Aspects of Sound Do the Sensory Receptors of the Ear Transduce?

Answer The three aspects of sound that is heard, interpreted and appreciated as transduced by the sensory hair cell receptors in the inner ear are pitch or tone, intensity or loudness and the timbre or quality of the sounds.

Concept Sound waves are travelling vibrations of air that comprise of high pressure regions of compressed air molecules alternating with low pressure regions of rarefaction of air molecules. Any instrument that makes sound or music produces such a pattern of disturbance in the air molecules.

The pitch or pure tone of a sound is determined by the frequency of vibrations. A higher tone or pitch of sound is produced by a greater frequency of vibrations. The human ear is able to detect sound waves ranging from 20 cycles per sec (Hz) to 20,000 Hz. Our ears are however most sensitive to sound frequencies in the 2000–3000 Hz range. Providentially, the ear is designed to hear conversations or normal speech that are within this region of sound frequencies.

The intensity of sound depends on the amplitude of sound waves or the pressure differences between the high pressure region of compressed and low pressure region of rarefied air molecules. Loudness of sound is quantified as decibels (dB) which is a logarithmic measure of sound pressure relative to a reference pressure in auditory physiology, near to the average hearing threshold at 1000 Hz i.e. $dB = 20\ Log_{10}$ Sound pressure/Ref pressure.

Thus if a sound pressure is 10 times that of the reference pressure, this will give a 20 dB. A sound pressure 1000 times that of the reference pressure will be from a 60 dB sound. Sound intensities greater than 100 dB is harmful to the auditory apparatus and pain is felt at close to 140 dB.

The timbre or quality of a sound depends on overtones or the additional sound frequencies that is superimposed on the fundamental pitch or tone. A tuning fork has a pure tone but different musical instruments playing the same C note sound nicely different and harmonize into the orchestral music that we enjoy.

Localization of sound is served by having two ears. The pinna of the ear reduces sound waves originating from the rear of a person and this changes the timbre of the sound and helps to discern the source of the sound. Sound waves reaches the ear nearer the sound source earlier than the farther ear. Also, the head absorbs or is a sound barrier and the intensity of the sound waves transmitted to the farther ear is decreased. The auditory cortex combines all these sensory information, including the timing pattern of neuronal firing to pinpoint and localize the sound we hear.

6 How Do We Perceive and Enjoy Soft, Soothing, Cello Music During an Evening Meal?

Answer The softness or loudness of the music is coded by frequency of action potentials from the auditory hair cells and the timbre of the cello unique tone is perceived when hair cells on different sites on the basilar membrane of the Cochlea respond to the mixture of sound frequencies that is produced by the cellist.

Concept The tympanic membrane vibrates in concert with the sound waves in the external ear. Then the middle ear bones (the ossicles malleus, incus and stapes) convey the tympanic vibrations into fluid movements in the inner ear when the oval window that is bridged by the ossicles to the tympanum vibrates accordingly. (note: the ossicular sound conduction is not what is defined as 'bone conduction' in clinical auditory testing).

Because of the greater pressure needed to move fluid in the cochlea, impedance matching is achieved first by the lever action of the ossicles. The ossicles also amplifies the pressure of the air-carried sound waves in the external ear to the cochlear fluid since the area of the stapes footplate on the oval window is much smaller than the tympanum. About a 20 times increase in force/pressure of the sound waves is transmitted.

The hair cells in the organ of Corti on the basilar membrane which forms the floor of the Cochlear duct transduce the amplitude and frequency of the fluid vibrations. There are around 15,000 hair cells within each cochlear and subdivided functionally into inner hair cells and outer hair cells.

The inner hair cells are where the receptor potentials are generated by the cochlear fluid vibrations that in turn produce oscillations in the basilar membrane. Mechanically-gated cation channels open to allow a unique potassium depolarizing current since the endolymph that bathes the stereocilia on each receptor hair cells is potassium rich and has a higher K^+ concentration than that inside the hair cells. The depolarized hair cells has voltage-gated calcium channels and the calcium influx triggers exocytosis of neurotransmitter (glutamate) onto the afferent fiber that innervates the hair cell.

The Outer hair cells do not transduce sound but exhibit electromotility; they shorten with depolarization and lengthen when hyperpolarized. This response somehow enhances the response of the sensory transducer inner hair cells to the amplitude and frequency of sounds.

The hair cells, especially the outer hair cells have also efferent motor innervations from the cranial VIII vestibulo-cochlear nerve.

The intensity of sound is determined by the amplitude of sound waves. A higher amplitude sound will lead to a greater basilar membrane oscillation at the peak responsive region (sound frequency -dependent). A greater bending of the stereocila results in a higher frequency of action potentials in the auditory sensory fiber.

Pitch discrimination is due to the properties and shape of the basilar membrane. The membrane is stiff and narrow at the oval window and wide and flexible at the helicotrema. High frequency pitches vibrate selectively the narrow part of the membrane and low bass notes deforms the membrane more towards the helicotrema. Each hair cell is interestingly tuned and activated best by its own adequate stimulus of a specific sound frequency. The sensory structure of Corti could thus be viewed as an Organ with 15,000 keys!

A pleasant timbre of a cello tone would thus stimulate via the fundamental frequency of the note and the additional frequencies of the overtones, the particular points on the basilar membrane. The pattern of frequency-specific hair cell stimulation and the afferent signals are then sent to and interpreted by the auditory cortex.

The primary auditory cortex in the temporal lobe is also tonotopically mapped. Each region of the basilar membrane is linked to a definite region of the primary auditory cortex.

7 How Do Our Eyes Focus Incoming Light Rays to Form a Clear Image on the Retina?

Answer Convergence of light rays to bring them into focus at the retina involves the two convex refractive surfaces of our eyes, namely the air-cornea interface and the lens.

Concept Light is a type of electromagnetic radiation consisting of photons that are transmitted in a wave-like manner. The wavelength is the distance between two wave peaks. The whole electromagnetic spectrum has an extreme wide range of wavelengths, from radio waves (10^4 m) to cosmic rays (10^{-14} m). The visible segment that is transduced and perceived as different color sensations fall within a very narrow range of 400–700 nm (10^{-9} m).

Convex surfaces converge light rays from any point source of light. In the eye the corneal-air convex interface is the major refractive site to bring light rays into focus. The refractive power at the cornea is however fixed but the refractive ability of the lens is changeable.

The Accommodation reflex increases the convexity of the lens and function to converge divergent light rays originating at objects close to the eye during near vision. The autonomic nervous system controls this reflex to refract light. Specifically, parasympathetic stimulation contracts the ciliary muscles that then leads to a slack in the suspensory ligaments that hold the lens. The recoil of the natural elasticity of the lens results in a shape change to a more convex refractive lens.

Age-related reduction in the lens elasticity will decrease the accommodative adaptability of the eye. This is called presbyopia and requires corrective lenses for near vision.

In myopia, near-sightedness or shortsightedness, the eye ball is apparently too long and a clear focal point on the retina is not achieved when light coming from a distant object is refracted. The image seen as blurred is focused in front of the retina.

The person however sees near objects clearly since the rays from a close distance are divergent before they are refracted by the eye.

The corrective lens will need to be concave to diverge the light rays before they enter the eye.

With the corrective spectacles, image of far objects are now focused onto the retina and near vision is also good with the usual accommodative reflex.

In far-sightedness, or long-sightedness (hyperopia), the eyeball is too short and thus the light rays from both far (parallel) and near objects (divergent) are both, in the absence of any reflex compensation, brought to focus behind the retinal layer. With accommodation reflex (which with constant use strains the eye), light rays from distant objects can still be clearly focused on the retina. Near vision is still blur/blurred even with accommodation to increase the convexity and refraction by the lens as the focal point is still behind the retina in the short eyeball.

A need to empower the refractive function of the eye is met by wearing spectacles with convex lens. This will enable distant view to be focused onto the retina without accommodation. Near vision is then recovered with the dual action of the convex lens and the accommodation reflex.

If the curvature of the cornea or lens is not uniform in all directions, astigmatism is the visual problem, as the refractive power is different in different planes causing part of the retinal image to be blurr. The optical correction is by cylindrical lenses to equalize the refraction in all meridians.

8 What Aspects of Receptor Potential Generation Is Unique in Photoreceptors?

Answer The receptor potential in photo receptors is due to a hyperpolarizing effect resulting from a reduced sodium cation current.

Concept The photoreceptors are rods and cones, both containing photo-sensitive pigments. The pigment in rods is rhodopsin that is made up of an aldehyde of vitamin A (11-cis-retinal) and a protein called opsin. The cones also have the 11-cis-retinal but the aldehyde is conjugated to different opsins which account for the specific light wavelength sensitivities.

Photoreception describes how light energy is trasnduced into electrical potential in the photo receptors.

In the photoreceptors, intracellular cyclic GMP (cGMP) regulates the permeability of the sodium channels. In the absence of light stimulation, there is a 'dark' current which is a due to sodium cation influx. The membrane potential of photoreceptors in the dark is low at about −40 mV.

When light strikes the photoreceptor, the rhodopsin is photo-isomerized and trans-rhodopsin is formed. A membrane G protein called transducin is activated which stimulates the enzyme phosphodiesterase that breaks down cGMP. The eventual reduction in photoreceptor cytosolic cGMP closes the ligand-gated sodium channels. A hyperpolarized receptor potential is generated in either the rods or the cones.

Rejuvenated rods (from trans to cis-retinal) is the basis of dark adaptation and this re-conversion process of the photopigment rohdopsin requires vitamin A, the deficiency of which leads to poor night vision.

The photoreceptors release the neurotransmitter glutamate. The hyperpolarized photoreceptor will secrete less glutamate onto bipolar cells. Bipolar cells are functionally separated into the ON-pathway ganglion cells and the OFF-pathway ganglion cells. The first action potential is produced at the retinal ganglions.

The basis of the ON or OFF pathways is due to the differential effect of the neurotransmitter glutamate on either metabotropic or ionotropic receptors on the bipolar cells. Metabotropic receptor binding hyperpolarizes bipolar cell. Thus when light stimulates this pathway, less glutamate from the hyperpolarized photoreceptor means this bipolar hyperpolarizing inhibition will be reduced. The bipolar cells can become more depolarized with release of excitatory neurotransmitter onto the retinal ganglion cells to excite a propagated action potential.

Conversely, in the OFF pathway, ionotropic receptors on the bipolar cells are bound by glutamate. This opens non selective cation channels, and a net depolarizing current occurs. When the photoreceptor is activated, less glutamate release will mean less depolarizing action on the bipolar cells will take place. A hyper-polarized OFF signal is the result.

Each photo receptor synapses with two such bipolar cells, also termed ON-center and OFF-center bipolar cells which are linked respectively to ON (Yes action potential) and OFF (No action potential) retinal ganglion cells.

9 How Is the Beauty of an Orange Sunset over a Green Garden Sensed?

Answer One major mechanism for color vision is based on the ratios of stimulation of the three cone types that are particularly sensitive to different peak wavelengths of light.

Concept Objects absorb selectively certain wavelengths of light and unabsorbed wavelengths are then reflected from the objects' surface. A blue object thus absorbs the longer green and red wavelengths of light and the shorter blue wavelength is reflected and activates the specific cone photoreceptors.

There are four different photo pigments, one in the rods and one in each of the red, green and blue cone photoreceptors. The absorption curves for the 3 cones overlap and this means that two or three cones can respond to stimulation by a single wavelength light but are activated to different degrees.

The absorption maximum for the four photoreceptors are 420 nm (blue), 530 nm (green), 560 nm (red) and 496 nm (rods).

For example, the sensation of yellow is perceived when the ratio of stimulation of the cones is 83:83:0, with red and green cones being equally activated at 83 % maximal and the blue cone is unaffected. White light is perceived when equal intensity of stimulation occurs to all the three types of cones. This trichromatic theory of color vision was formulated by Thomas Young.

In addition, a more recent mechanistic concept for color vision involves what is called opponent color cell theory. Proposed by E Hering, specific ganglion cells are said to code for specific colors. The 'opponent' refers to the fact that excitatory input and inhibitory input are received from different cone types. Blue and yellow are opponent colors and so are green and red. There is evidence that some retinal ganglion cells are excited by red light at the center of their receptive fields but are inhibited by green light in the surround area.

The rhodopsin in every rod, when stimulated by a given wavelength will respond identically. As such, no discrimination between light wavelengths is provided by the sensory inputs from the rod photoreceptors. Therefore rods, although more sensitive than cones allow night vision in dim light but the images seen are different shades of gray since light intensity and not color is transduced.

The gene for blue sensitive cone pigment is on chromosome 7 but the red and green pigments are encoded on the X chromosome. If one cone type is functionally deficient, the person's color blindness will be defined by the differential response of the remaining two cone photoreceptors to different light wavelengths.

10 How Are Visual Sensitivity and Visual Acuity Dependent on the Neurophysiology of the Photoreceptors?

Answer Rod photoreceptors are more sensitive but transduce poorer visual acuity than the cone photoreceptors due mainly to the different degrees of neural convergence between the rods and cones.

Concept There are 20 times more rods (120 million) than cones (6 million) in the retina. Cones are most abundant in the macula lutea in the center of the retina. Spreading out from this point the concentration of cones decreases while the population of rods increases.

The higher sensitivity of the rods is not simply due to their greater number. Structurally, the outer segments of the rods are longer and contain more photopigments which will absorb more light.

Neural convergence explains the better sensitivity of the rods. The retina has 1 million retinal ganglion cells which receive inputs from the 126 million photo receptors. Thus on average, the sensory afferents from 100 rods via bipolar cells converge onto a single ganglion. Since the first action potential is triggered at the ganglion cells, the convergence provides the summative potential to reach the firing threshold quickly.

In contrast, the retinal cones have frequently a one-on-one linkage to the associated ganglion cell. As such, the conal sensitivity is less than for the rods which are used for vision in dim light. However, since the retinal receptive field for the cone photoreceptors are very small, the sensory visual acuity served by the cones is high. For the rods, visual sensitivity is achieved at the expense of visual acuity. The image transmitted by rods are not sharp as for the cones that are operative in colorful day vision.

Light adaptation occurs on the basis of the different photoreceptor sensitivities. When emerging from a dark cinema hall, the light in the concourse dazzles because the sensitive rods were functionally revived during the film show. A rapid reduction in photoreceptor sensitivity soon enables the cinemagoer to see comfortably again. The rod photoreceptors are bleached by the light and optimal vision is resumed by reverting to the less sensitive cones for the light transduction.

Architecturally, light entering the eye quite unusually must pass through the ganglion and bipolar cell layers before reaching the retinal photoreceptors, except at the fovea. At the fovea, the ganglion and the bipolar layers are pushed aside and light hits the foveal cone photo receptors directly. The foveal acuity is the highest in the retina.

Patients with macular degeneration have poor visual acuity in the middle of their visual field. Their remaining peripheral vision which are served by the rods that have less sensory acuity is also poor and indistinct.

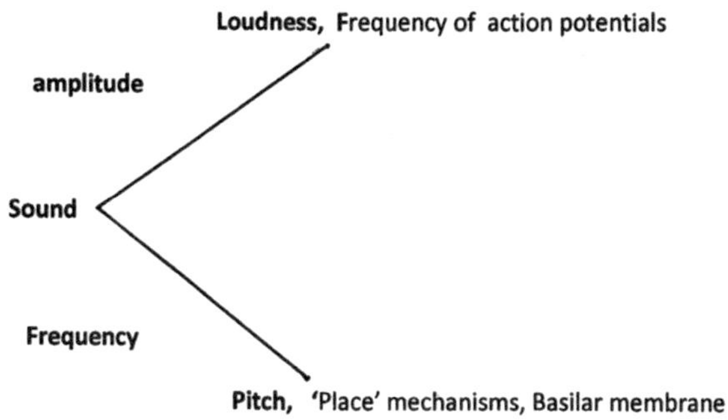

Sound waves are decoded in relation to their amplitude and frequency. The amplitude of sound waves determines the loudness of sound heard which vary with the frequency of the transmitted action potentials, AP (note: amplitude of AP, unlike the sound waves is constant). The frequency of the sound waves gives the pitch of the sound and this is discriminated by the site-specific, peak vibrations of the basilar membrane that holds the hair cells in the cochlear.

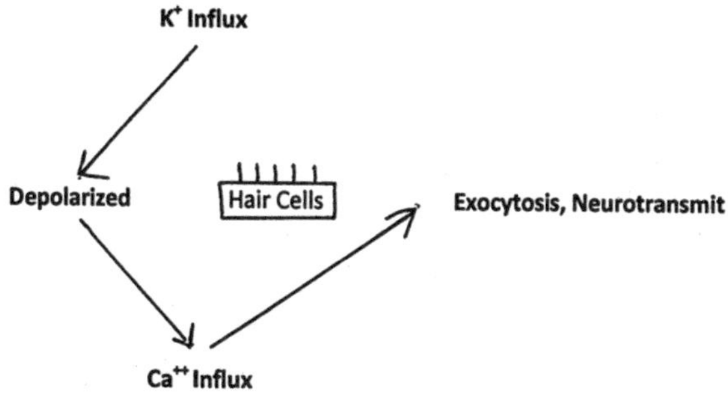

The hair cells are unique in their ionic basis of membrane depolarization. An inward potassium cation current depolarizes the cell instead of the usual sodium current influx. ECF calcium then enters via voltage-gated channels and triggers exocytosis of neurotransmitter onto auditory or vestibular afferent nerve fibers.

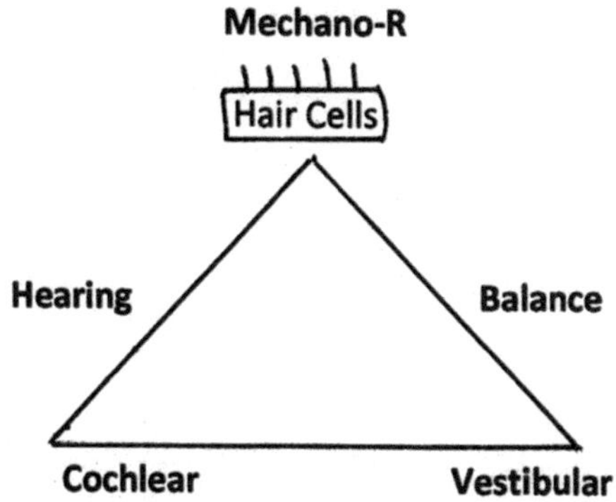

The sensory transduction and perception of hearing and the sense of balance and equilibrium both involve mechano-receptors. These are hair cells in the auditory cochlear and vestibular apparatus respectively. Sound energy is transmitted through air and water media to bend cochlear hair cell receptors. Fluid movement in the vestibular semicircular canals contorts hair cells that detect rotational acceleration. Gravitational force that produces linear acceleration affects the hair cells of the otolith organs.

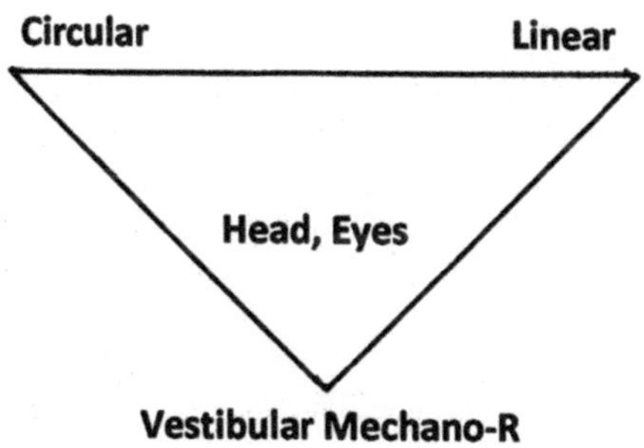

The vestibular transduction mechanisms serve to provide information to the CNS on the position of the head in space. The vestibulo-ocular reflexes function to maintain the image focus on the cone-rich area of the retina during head movements.

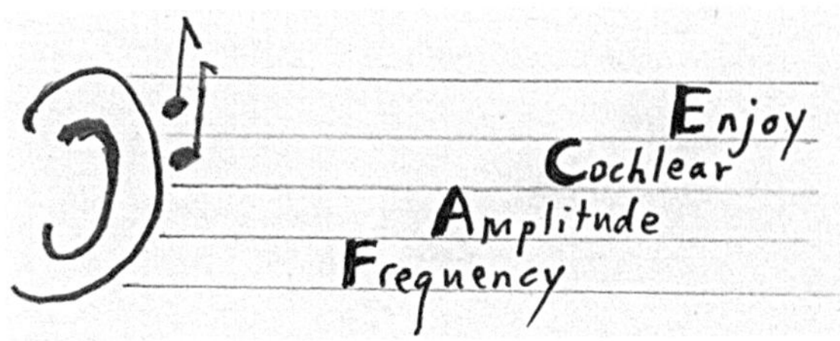

The amplitude and the frequency of the sound waves determine the loudness and pitch of the sound. The activity and location of hair cells in the Cochlear transduced the mechanical pressure of the sound energy to meaningful sounds or music that we enjoy. There is in addition, involvement of language and mood centers in the brain when the music is sung and appreciated.

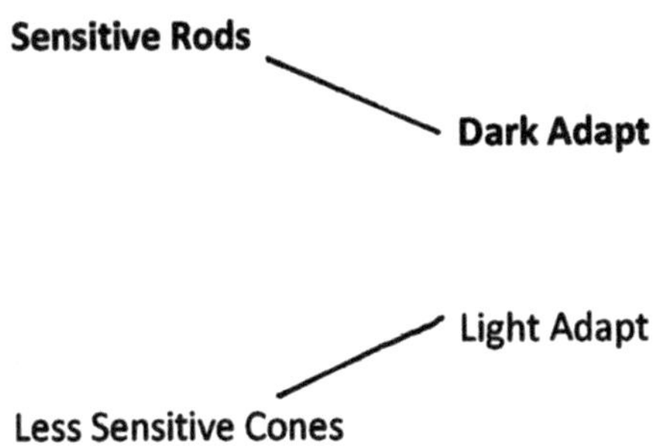

The two groups of retinal photoreceptors have different sensitivity and acuity. Cones are less sensitive than rods but their sensory transduction produces a clear image. In contrast, the more sensitive rods are used to see less clearly in dim light.

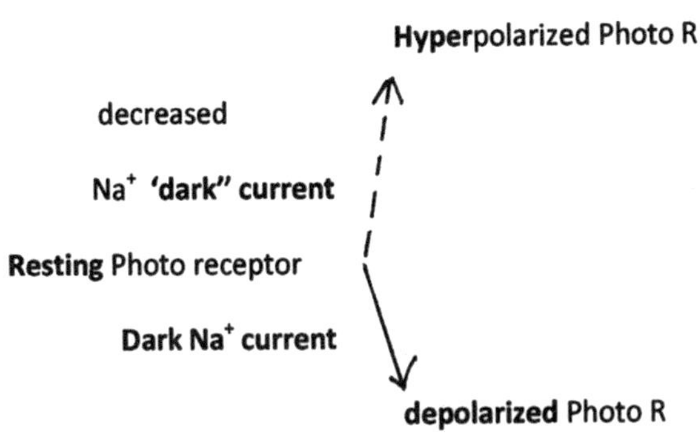

The retinal photo receptors are unusual in their sensory transduction response. Photoreceptors become hyper-polarized when stimulated by light. Resting photoreceptors are partially depolarized by a sodium current. Inhibition of this sodium cationic current during photo-transduction leads to a more negative membrane potential. This hyperpolarized response is however eventually converted through the multi-cellular retinal layers to generate an action potential in the retinal ganglion.

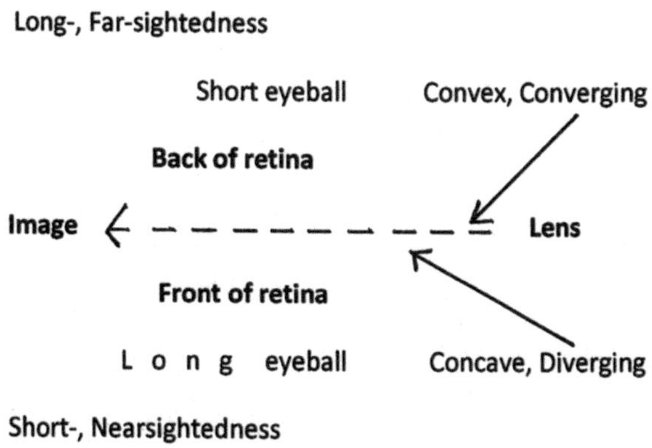

Problems in light refraction lead to the images being focused either in front or behind the retina. The cornea-retina length of the eyeball accounts for the long or short-sightedness. Corrective lenses converges or diverges incoming light respectively so that a retinal image is restored.

Chapter 5
Muscles and Movement

Living things move. Physiology is a moving subject. Even when we are at rest, there are a variety of dynamic movements in our body. Blood is flowing, air is inspired and expired, peristalsis of chewed food and chyme during a meal and in the nephrons, tubular/urine flow. These internal flows are achieved mostly by smooth muscle and cardiac muscle function that are regulated by the autonomic nervous system (ANS). Secretory movement of exocrine digestive juices and endocrine secretions are also the result of ANS actions.

The somatic nervous system co-ordinates all voluntary, intentional movements. The brain plans, initiates the motor activity and the corresponding muscle groups are stimulated to contract, with concurrent relaxation of the antagonistic muscles. The final activated neurons to the respective muscles are either brain stem or spinal alpha motor neurons that govern skeletal muscles in the head/neck region and the rest of the body respectively.

Skeletal muscle movement is uniquely fine tuned by sensory feedback proprioceptive signals from the muscle spindles, tendon and joint receptors. The cardiac and smooth muscles basically just respond to the effector ANS. These skeletal muscle proprioception are transmitted to the cerebellum and other higher, cortical motor centers that are involved in effecting a focused, directed movement that matches the original thought.

Writing or typing these words requires the co-ordinated fine motor movements of my digits (digital control by electrical action potentials along specific alpha motor neurons!). The ability to enjoy conversations with friends is possible by the orchestrated movement of our tongue and speech muscles, which in turn are intimately associated with listening and understanding the language of communication.

H.M. Cheng (ed.), *Physiology Question-Based Learning*,
DOI 10.1007/978-981-10-0877-1_5

1 How Is the Primary Motor Cortex Involved When You Take a Step over a Drain?

Answer The primary motor cortex is not the initiation site of motor muscle movements. Three major higher motor functional areas of the cortex provide the command instructions to the primary motor cortex namely, the pre-motor cortex, the supplementary motor area and the posterior parietal cortex.

Concept Electrical activity have been observed in areas around the primary motor cortex, about 750 ms before activation of the primary motor cortex that controls muscle movements. This electrical discharge has been called readiness potential and is generated at the highest level of a functional hierarchy of neurons that control body movements.

The middle hierarchical structures would include the collective area termed sensorimotor cortex, sub-cortical nuclei that includes the basal ganglia and thalamus, the cerebellum and brainstem nuclei. Much of the middle level regulatory neurons in motor control ensure the appropriate postures during movements and achieve focused accurate movements as intended.

The middle level motor neurons not only receive instructional signals from the highest command neurons. Afferent sensory information are continually received, sent from the proprioceptors (muscles, tendons, joints, skin), the vestibular apparatus and the visual neurons. The 'middle kingdom' neurons integrate the command, descending and afferent, ascending signals to effect the desired motor program of co-ordinated muscle contraction/relaxation.

The final common pathway, the target neuron is the alpha motor neuron that innervates the specific muscles. The cortico-spinal descending pathways (sometimes called pyramidal tracts/system) terminate at the spinal alpha motor neurons that serve muscles in the body except the head and neck. The cortico-spinal neurons control fine, isolated movements, in particular those by the hands and fingers.

The cortico-bulbar pathway that begins at the sensorimotor cortex and reaches the brainstem alpha motor neurons, control muscles of the eyes, face, tongue, throat.

Axons that arise from the brainstem also impinge and influence the activity of spinal cord alpha motor neurons. These so called 'extra-pyramidal' tracts include the vestibulo-spinal and the reticulo-spinal pathways from the vestibular and reticular formation nuclei in the brain stem respectively. The brainstem pathways are essential in the maintenance of upright posture, balance and during 'unconscious, involuntary' walking. The brainstem neurons target large muscle groups in the neck, trunk and upper limb portions.

All movements whether voluntary or 'automatic' involve interactions between the cortico-spinal and brainstem descending pathways. Use of the term 'involuntary' is quite easily confused and does not reflect the normal experience of body movement physiology. 'Involuntary' movement is assumed to be synonymous with movement of an 'unconscious, automatic and reflex' nature.

Note that the word 'reflex' is used in 'voluntary' skeletal muscle physiology e.g. 'muscle spindle reflex'.

All motor muscle responses include conscious and unconscious neural activities. The maintenance of posture while you are reading this note on muscle neurophysiology is 'unconscious'. The contributing degree of conscious/unconscious or voluntary/involuntary actions changes with the frequency of repeating the specific motor movements. Think of a child practicing her musical scales on the piano, from initial deliberate, determined effort to remember the fingering to eventual almost 'automatic' fluid execution of playing the ascending and descending scales, up and down the black and white piano keys.

2 During Movement, What Sensory Information from the Muscles Are Fed Back to the Central Nervous System?

Answer As the muscles contract or relax, information about static and dynamic muscle length, muscle tension are fed back to CNS to monitor and ensure that the movements taking place are executed as planned in the higher motor centers.

Concept The alpha motor neurons in the spinal cord and brain stem (the latter exiting within some cranial nerves) are the final effectors of specific muscle contractions during any movement. Note that in isometric contractions, there is no physical movement of the muscles in space.

Most of the synaptic inputs from descending pathways and afferent neurons onto motor neurons are via interneurons. These interneurons that make connections with the alpha motor neurons are either excitatory or inhibitory neurons.

As the muscles move, local reflex circuits are continually activated by changes in the length and tension of the muscles. The intrafusal muscle spindles and the Golgi tendon organ convey the afferent signals for muscle length and tension respectively. The sensory receptors at the overlying skin where the muscles moves and the joints also send proprioceptive information, that together with signals from the spindle and tendon receptors to the cerebellum, brain stem and subcortical nuclei refine and fine tune the ongoing movements. Afferent information that contributes to the conscious appreciation of the body position and its moving parts in space is relayed continually.

Besides the ascending pathways of the afferent impulses to the higher hierarchical functional neurons of motor control, the afferent fibers also synapse with the alpha motor neurons via the interneurons or directly. The muscle spindle reflex is a mono-synaptic reflex that excites the alpha motor neuron supplying the extrafusal contracting muscles.

The Golgi tendon organ receptors when activated will excite an inhibitory interneuron that then synapses with the alpha motor neuron that serves the same muscle. The antagonistic muscle during a Golgi tendon reflex will be stimulated to contract by an excitatory interneuron.

When we consider a body movement, the antagonistic muscle (e.g. extensor) is relaxed when the flexor muscle is contracted. The alpha motor neuron to the extensor is thus inhibited by an inhibitory interneuron that is stimulated by a branch of the activated afferent fiber arising from the muscle spindle receptors in the flexors. This interneuronal involvement is part of what is termed reciprocal innervation. Contraction of synergistic muscles involved in e.g. flexing the arm will be achieved by simultaneous excitation of synergistic alpha motor neurons.

Thus the local afferent inputs to the alpha motor neurons come from three directions; (i) from muscle spindles in the skeletal muscles innervated by that motor neuron (ii) from other muscles e.g. antagonistic paired (iii) from sensory receptors in tendons, joints and the involved skin.

3 How Are the Efferent Innervations to the Muscle Spindles Involved in Voluntary Actions?

Answer The gama efferent nerve fibers to the intrafusal muscle spindles are coactivated with the associated alpha motor neurons during voluntary movement of the same muscles.

Concept When voluntary changes in muscle length are triggered at motor areas of the cerebral cortex, the neuronal command simultaneously activates both the alpha motor neurons and the associated gamma motor neurons in the spinal cord or brain stem (for muscles of head and neck).

In the classic knee-jerk reflex, stretching of the quadriceps muscle by the patella tendon tap activates the muscle spindles that are located in parallel to the extrafusal muscles. A contraction reflex resulting from synchronous discharge of alpha motor neurons supplying the quadriceps muscle produces the knee jerk.

When muscles contract, the muscle spindles slacken. The reduction in spindle tension will decrease the responsiveness of the muscle spindles to relay sensory information about muscle length. The gamma efferent fibers innervate the two ends of the intrafusal muscle fibers. As such, when the gamma motor neurons are concurrently activated during a motor command from the cerebral cortex, the muscle spindles do not slacken with the muscle contraction. The sensory spindles are kept in tension by the gamma efferent activity and the spindle sensitivity is maintained throughout the muscle activity so that afferent information about changing muscle length continues to be fed back to the CNS.

The axons of the gamma motorneurons are also known as fusimotor fibers. The fusimotor fibers innervate both the two types of intrafusal muscle spindle fibers, the nuclear bag and nuclear chain. The contractile ends of the intrafusal muscles, when activated, are not strong to cause shortening of the extrafusal muscle fibers.

The muscle spindles are innervated by two types of afferent fibers, Ia and II. Afferent Ia fibers are dynamic endings and monitor rapid changes in length while sensory II fibers serve as static muscle length comparators.

Skeletal muscle tone is due to the muscle spindle contraction reflex produced by passive muscle stretch. The normal look of a face with unopened mouth is due to the muscle tone activated from a spindle reflex that holds our jaw up.

Increased gamma efferent activity could then potentially increase spindle responsiveness and lead to a hyperreflexia and hypertonic muscle.

Clinically though, hypertonia is commonly associated with what is classified as 'upper motor neuron' dysfunction. This is caused by the loss of descending pathways that provide the physiologic inhibitory modulation of alpha motor neuron activity.

4 How Is the Contraction of Whole Muscle Graded or Increased In Vivo?

Answer Since whole muscles are made up of muscle fibers distributed into motor units, scaling of whole muscle contraction is via recruitment of the motor units and also by tetanic summation.

Concept All muscle fibers in a single motor unit are of the same fiber type, either slow-oxidative (Type I), fast-glycolytic (Type IIb) or fast-oxidative-glycolytic (Type IIa). Students should note not to confuse the Muscle fiber types with the sensory nerve fiber type (I to IV). So functionally, or metabolically, the motor units can be called Type I, IIa, IIb motor units.

All the motor neurons that supply a given muscle is collectively a motor neuron pool. The cell bodies of the neuronal pool are aggregated near each other either in the ventral horn of the spinal cord or in the brainstem. The axons of the motor neurons exit in the spinal nerves or the cranial nerves. Definitions and language affect our learning. By strict definition, a nerve is a bundle of peripheral axons, efferent motor or afferent sensory, encased by connective tissues. Since a nerve, thus defined does not include the whole nerve cell (neuron), there are no nerves in the CNS! (bundles of axons in the CNS are termed tracts).

Alpha motor neuron recruitments are dependent on the size of the motor neurons. The diameter of the motor neuron cell body correlates generally with the axonal diameter. A single excitatory synaptic input will cause the same amount of sodium influx or current on either a small or larger motor neuron. The smaller neuron will be depolarized more due to the smaller cell membrane area. Thus in a motor neuron pool, the smaller neurons are recruited first and action potentials will be generated earlier to the motor units that they supply.

The size of the motor units increases from slow-oxidative to fast-oxidative-glycolytic (fog) to fast- glycolytic (fg) fiber type motor units. Thus slow-oxidative motor units are recruited earliest in a whole muscle. The proportion of the three motor unit types depends on the muscle function. For the back muscles which must sustain contraction for long lengths of time to maintain upright posture, there is a greater number of slow-oxidative fibers which are characterized by low fatigability.

Training and required muscle function contribute to the degree of motor unit representation in the muscles. For example, sprinters have more than 70 % fast-glycolytic motor units in their gastronemius muscle of the lower leg. In contrast, in distance runners, the same muscle can have above 70 % of slow fiber motor units.

When all the motor neurons in a pool are recruited, the maximal tension of the whole muscle is still not reached. Increasing the frequency of action potentials transmitted via the alpha motor neurons by a voluntary cerebral effort will produce tetanic summation since the mechanical contraction of muscles do not have a refractory period, unlike in the electrical activity of an action potential. Up to a five-fold increase in muscle tension above a single motor unit twitch results from a tetanic summation.

Motor neurons and sensory neurons are also defined respectively as the final common efferent pathway to the target organs and the innervating afferent nerve fibers from the location of the sensory receptors (separate receptors or the nerve terminals themselves). Here again, students should distinguish between the labeled, functional sensory/motor areas of the cortex and the peripheral motor/sensory neurons, whose axons travel in the spinal or cranial nerves. Marieb makes the insightful physiologic statement that all neurons in the cortex are interneurons.

5 How Do the Metabolism of Muscle Types Determine Their Mechanical Functions?

Answer The major biochemical pathway for production of ATP in different muscle fiber types determine their maximal velocities of shortening.

Concept Fast twitch and slow twitch fibers contain different forms of myosin ATPase that hydrolyse ATP at different maximal rates. Fibers with low ATPase activity are called type I (slow-oxidative I fibers). Students should note that for sensory fibers, type I has the fastest conducting velocity for electrical impulses. Fast muscle fibers are type II fibers. To bring all these rather confusing numbering together, we can say that all muscle fiber types (I, II) are innervated by afferent sensory fibers (types I, II) that monitor the dynamic/static length of the muscle spindles.

The oxidative fibers have many mitochondria and thus the ATP production is dependent on adequate blood supply. The muscle fiber capillarization is therefore more dense. The myoglobin content of slow fibers are high and this oxygen storage protein has a high affinity for oxygen. Myoglobin in slow oxidative fibers account for the red coloration of these 'red meat' muscle fibers.

Oxidative fibers have lower rate of fatigue compared to fast-glycolytic fibers. The fast fibers however generally have larger diameters and thus possess a higher number of thick/thin filaments cross bridging points per unit cross sectional fiber area to generate a greater force or tension during contraction. Fast fibers lack mitochondria but has a high concentration of glycolytic enzymes and glycogen content.

The contraction velocity or shortening velocity of fast fibers as their name implies is rapid. Generation of large muscle tension over a short period is thus a property and function of fast fibers. The shortening velocity of the whole muscle is also affected by the load on the muscle. The shortening velocity is zero when the load is equal to the tension produced by the maximal isometric contraction.

Isotonic contraction occurs when the load is moved. The power of a muscle is equal to the product of the load and the velocity of load movement. Recruitment of motor units not only increases the muscle tension but can also increase the velocity of muscle shortening. At a constant load, activation of more motor units will move the load at a higher velocity.

A fast-glycolytic motor unit generally has more innervated motor fibers compared with a slow-oxidative motor unit. Combined with the larger diameter of fast fibers, activating a fast motor unit will generate a greater force than recruiting a slow motor unit.

Muscle tension is generated within the sarcomeres, the contractile component of the muscle by the sliding of thin and thick filaments. The sarcomeres do not connect directly to the bones. The tension generated by the sarcomere shortening is transmitted to the bone through the connective tissue and tendons. These, together with the intracellular giant protein titin, form what is called the series-elastic component of the muscles. The tightening of this series elastic element during sarcomere contraction moves the bone against the load.

6 How Is Energy Used in the Resting Muscle and Exercising Muscle?

Answer In the resting muscle cell, ATP is used to maintain trans-membrane chemical gradient. When skeletal muscles are stimulated by action potentials, a sequential activation of myosin ATPase and calcium ATPase occur during contraction and relaxation respectively.

Concept The skeletal muscle fiber requires energy like all living cells, using ATP as the sole energy source. The membrane ATPase activity maintains the trans-membrane concentration gradients for both sodium and potassium. These cation gradients are exploited when the alpha motor neurons stimulate the muscle. Excitation-contraction coupling mechanisms generate action potentials in the muscle membrane via acute sodium current and potassium current that flow through their respective voltage-gated cation channels.

At the beginning of a muscle contraction, intracellular creatinine phosphate provides the first readily available energy stores. Creatine phosphate donates a high energy phosphate to ADP to form ATP. Depending on the muscle types, the other two pathways of ATP production are oxidative phosphorylation and glycolysis.

During the cycle of contractile mechanisms when cross bridges between the actin and myosin filaments interact, ATP is needed. The myosin ATPase splits ATP to energize the power stroke of the cross bridge during the filament sliding event that shortens the sarcomeres.

In addition, the binding of an unsplit ATP to myosin during the cross bridge cycling is required to detach the myosin head of the thick filaments (thick headed!) from the thin actin filaments. The stiff muscles of a dead body, depleted of ATP highlights this releasing function of ATP.

Action potentials in muscle depolarize the myocyte fibers that lead to an increase in cytosolic calcium, released from the sarcoplasmic reticulum. Calcium binds to troponin and unleashes the sarcomere shortening. The relaxation of muscles involves the reversal of calcium flux into the sarcoplasmic reticular calcium stores. The lowering of intracellular calcium in the muscle fibers is driven by the sarcoplasmic calcium ATPase.

Muscle performance and muscle fatigue can be due to a variety of factors that affect the transmission of signals from the cerebral cortex via the alpha motor neurons to the target muscles. Neuromuscular junction fatigue is the situation if the synthesis and release of acetylcholine neurotransmiiter cannot keep up or sustain the continuous arrival of action potentials at the terminals of the motor neurons.

Locally, depletion of glycogen, a main source of fast-glycolytic muscles also contributes to muscle fatigue. Accumulation of lactic acid could also inhibit essential enzymes in the ATP production and steps in the excitation-contraction coupling. Tissue hyperkalemia occurs due to the potassium efflux during each action potential. The alteration in membrane potential might result in a depolarization block (due to inactivation of sodium channels) at the T-tubules, which is necessary for triggering from the adjoining sarcoplasmic reticulum, the calcium release.

Central command fatigue describes the failure to send excitatory signals to the alpha motor neurons. The person stops the physical activity even though the muscles are neuromuscularly and biochemically not fatigued. Thus an athlete's performance is also influenced by the motivation and perseverance especially noted in strenuous competitions.

The muscles that make us smile are innervated from motor neurons in the brain stem. Tiredness, boredom and monotony do contribute to the absence of our normal smile 'reflex' when we meet another person. Remember, only humans smile!

7 What Aspects of Voluntary Movements Is the Cerebellum Essential For?

Answer The cerebellum generates an 'error signal' to the cerebral motor areas by comparing intended actions with actual, moment-by-moment muscle status via proprioreception to adjust and effect a precise and timed movement.

Concept Electrical stimulation of the cerebellum does not produce sensation or initiate movement. Unlike damage to sensorimotor cortical areas, lesions of the cerebellum are not usually associated with muscle weakness or paralysis.

From clinical correlation with cerebellar lesions, the cerebellum is essential in postural control and executing smooth rapid movements initiated by the cerebral cortex. Unstable posture and awkward, staggering gait (ataxia) are the presenting symptoms in cerebellar dysfunctions.

During movement, planned and initiated by the cerebral cortex, sensory signals are relayed to the cerebellum from muscle spindles, tendon organs, joints and cutaneous touch receptors by way of the spino-cerebellar tracts. For posture control and adjustment during movement, vestibulo-cerebellar signals are also transmitted.

These sensory afferents, besides neural information along the cortico-cerebellar and reticulo-cerebellar pathways enter the cerebellar cortex via the two major inroads, the climbing fibers and the mossy fibers. The Purkinje cells that represent the only efferent outlet from the cerebellar cortex are excited by both the climbing and mossy fibers (to remember, you could make a statement like "an exciting climb up a mossy path to Purkinje land"!).

The Purkinje neuron is an inhibitory neuron and releases GABA at its nerve terminals on deep cerebellar nuclei. These nuclei also receive excitatory synapses from collaterals of the mossy and climbing fibers. A net cerebellar neural signal output is thus modulated.

The terminal points of the spinocerebellar pathways are somatotopically represented in the cerebellum. Interestingly there are two such sensory maps in the cerebellum, an anterior and a posterior cerebellar 'homunculus'.

Cortico-cerebellar signals inform the cerebellum of an intended action. During the movement, afferent sensory signals monitor continually the progress of the muscle movement. The cerebellum compares the 'blueprint' with the real time movement and generates an 'error signal' if needed. The homeostatic error signal is fed back to the cerebral motor areas to fine-tune for a precise, smooth motor movement.

Cerebellar lesions are thus often associated with an intention tremor, in which the abnormal movement of the patient oscillates around a desired intended position and direction.

The cerebellum has no direct connections with the cerebral cortex and also no direct synaptic links with the spinal motorneurons. Unlike the cerebral cortex, the cerebellar hemispheres each control and receive inputs from muscles on the ipsilateral side of the body.

The cerebellum might also be involved in certain types of motor learning. It has been observed in monkeys that the frequency of action potentials in climbing fibers increases when the monkey has to learn a new task (remember 'climbing monkeys!).

8 How Many of the Twelve Cranial Nerves Control Muscle Function of the Head and Neck?

Answer Nine of the cranial nerves serve motor functions. The other three cranial nerves, olfactory (I), optic (II) and vestibulocochlear (VIII) are purely sensory nerves. All the nine cranial nerves that have motor nerve fibers also carry afferent sensory fibers ('mixed nerve') from proprioceptors in the muscles they control.

Concept The first two pairs of cranial nerves (olfactory, optic) are attached to the forebrain. The other ten cranial nerves all originate in the brainstem. The vagus nerve is a 'wanderer '('vagabond') and is the only cranial nerve to explore and extend beyond the head and neck region to fufill its neural roles. Almost all the motor fibers in the vagus nerve are parasympathetic efferents except for those somatic fibers that innervate skeletal muscles of the pharynx and larynx that are used in swallowing.

The vagus nerve together with the occulomotor (III), facial (VII) and glossopharyngeal (IX) contain the cranial parasympathetic neural outflow. The occulomotor parasympathetic fibers are involved in the accommodation reflex for near vision by its action on iris muscle contraction (pupil size) and ciliary muscle contraction (lens shape).

The occulomotor nerve also contains somatic fibers to four of the six extra-ocular muscles. The term 'mixed nerve' can thus be used to describe a nerve in which both autonomic and somatic fibers travel. The trochlear nerve (IV) and the abducens nerve (VI) contain somatic motor fibers that each control one of the extra-ocular skeletal muscles.

All the nerves that control the extrinsic eye muscles (occulomotor, trochlear, abducens) also carry afferent sensory fibers from proprioeceptors in the muscles that they serve. This paired motor/sensory anti-parallel neural highways are also present in the accessory (XI) and the hypoglossal (XII) cranial nerves.

The accessory nerve has motor fibers that include those that supply the trapezius and the sternocleidomastoid muscles that move the head and neck (to remember, think of the accessory necklace!). As the name suggests, the 'glossal' nerve essentially contributes to the tongue movements in swallowing and speech.

Life (food) will be tasteless without the function of sensory fibers in the glossopharyngeal, vagus and facial nerves. Motor fibers in the glossopharyngeal nerve (IX) are also involved in the swallowing and gag reflex. The parotid salivary glands are stimulated by parasympathetic motor fibers in the IX nerve (note the term 'motor' is not restricted to somatic, skeletal muscle movements; the salivary flow is stimulated to 'move'/flow).

Give the perfect Smile for the 7th time for facial nerve (VII) that enables us to project the whole spectrum of facial expressions.

The trigeminal nerve (V) is the largest of the cranial nerves. Three branch divisions of the nerve, ophthalmic, maxillary and mandibular convey afferent sensory impulses from touch, temperature and nociceptors. The motor fibers in the mandibular sector supply the muscles of mastication from which sensory feedback proprioceptors are transmiited in the same mandibular branch nerve. (Chew a triGem!)

The sensory roles of cranial nerves (vagus, glossopharyngeal) are crucial in the homeostatic regulation of blood pressure and breathing. Arterial blood pressure and arterial gas partial pressures are monitored by baroreceptors and chemoreceptors respectively. Afferent signals are sent from these arterial mechano- and chemosensors via the vagus and the glosspharyngeal nerves to the brain stem cardiorespiratory regulatory neuronal centers to effect the necessary compensatory effector responses. Viva La Vagus!

9 How Do the Somatic and Autonomic Peripheral Neural Pathways to Muscles Function Integratively to Keep Us Alive?

Answer The somatic alpha motor neurons when activated leads always and only to skeletal muscle excitation, including the respiratory diaphragm. The autonomic motor fibers innervate cardiac and smooth muscles which provide the rhythmic pump and vascular conduits respectively for the life-giving blood.

Concept The somatic peripheral motor division consists of a single alpha motor neuron between the central nervous system (spinal cord or brainstem) and skeletal muscle fibers. The transmitter at the neuromuscular junction, acetylcholine when bound to nicotininc receptors at the end-plate always depolarizes the muscles. Relaxation is not due to any inhibitory neurotransmission but involves cessation of action potential transmission along the alpha motor neuron to all the fibers in each motor unit (active sarcoplasmic calcium ATPase activity is also needed for muscle relaxation).

The normal breathing rate is achieved by pacemaker potential generation and transmission from the respiratory neurons in the brain stem to the inspiratory muscles. A ten action potentials/min cycle in the respiratory neurons will produce ten contraction/relaxation of the diaphragm, the major skeletal inspiratory muscle. The breathing rate will then be ten tidal volume/min.

The autonomic motor efferents supply the cardiac muscle and smooth muscle. The pathway is bi-synaptic between the CNS and the effector cardiac/smooth muscle. Neurotransmission in the autonomic fibers can either stimulate or inhibit muscle contraction.

The heart rate is under dual autonomic control. The basal cardiac pacemaker activity from the sino-atrial nodal cells is increased by noradrenergic sympathetic and reduced by cholinergic parasympathetic neural input.

The vascular resistance that regulates regional blood flow is served by arteriolar smooth muscles. The control of vascular resistance is basically via a mono-sympathetic vasoconstrictor action. Vaso-dilation to reduce vascular resistance and increase blood flow is mediated by a decrease in the sympathetic discharge to the arterioles (this neural effect occurs together with action of local metabolite vasodilators during physical activity).

Sympathetic action also venoconstricts the smooth muscles of the veins, the capacitance blood vessels that compliantly accommodate most of the total blood volume. Venous return is increased to provide a greater cardiac output ejected by the cardiac muscle pump into the closed circulatory system.

During exercise, contraction of the skeletal muscles by somatic motor neurons also enhances ventricular infilling by a 'muscle pump' mechanism to sustain a higher cardiac output. Increased activity of the inspiratory skeletal muscles, in addition effects a suction 'respiratory pump' to better the venous return.

Nutrition to sustain energy demands by the body is the role of gastrointestinal (GI) physiology. We masticate somatically. Autonomic motor efferents (predominantly parasympathetic) not only move the food peristaltically along the smooth muscle tract but also stimulates the required salivary, gastric and pancreatic digestive secretions. The enteric system of the GI tract ('brain of the gut', gut-feelings!) is thus linked to the CNS by the autonomic nervous pathways.

10 Why Is the Hypothalamus Described as a 'Head Ganglion' in the Autonomic Nervous Control of Cardiac and Smooth Muscle Response?

Answer The hypothalamus behaves like a chief ganglion at the top of the hierarchy of autonomic neural outflow to smooth muscles involved in e.g. urination, defecation, erection.

Concept The autonomic nervous system (ANS) should be viewed as including the visceral reflexes that determine the degree of sympathetic and parasympathetic output. The medulla in the brainstem contains neurons that are central to regulating cardiovascular, respiratory and digestive activities. Cardiac muscle and smooth muscle responses are part of the compensatory autonomic effector mechanisms during the visceral reflex that monitors blood pressure via baro-receptors. Note that in respiratory control, a visceral reflex by the arterial chemoreceptors is uniquely linked in the homeostatic response loop to skeletal, inspiratory muscle contraction.

The hypothalamus exerts its control of ANS directly or via synaptic connections at the reticular formation, which in turn influences pre-ganglionic motor neurons in the brain stem or spinal cord.

Through its association with the limbic system, the hypothalamus accounts for our autonomic reactions to emotional situations. Fear, anger e.g. are associated with changes in heart rate, blood pressure and breathing.

Cerebral cortical signals in turn can also modulate autonomic functions by its association with the limbic system. Embarrassing memories when we look at a photograph can make us blush which is produced by vasodilation of cutaneous blood vessels in our cheeks.

Defecation (poo) and micturition (pee) are spinal cord autonomic reflexes that, in adults are subject to appropriate conscious inhibition of smooth muscle functions in the rectum and bladder respectively. Strong emotional stimuli can via hypothalamic neurons cause e.g. incontinence or nervous diarrhea.

Sexual excitement produces penile erection involving limbo-hypothalamic stimulation. In turn, clinical problems of erectile dysfunction are also due to socio-psychological reasons.

From studies of persons who practice meditation, there is a potential use of biofeedback training in helping the persons to focus on calming, soothing thoughts. Desired changes in some physiological parameters are then audibly or visibly cued. These responses include heart rate, blood pressure, skeletal muscle tone that are altered by stress affecting the persons. Keep Calm and be Physiologic!

10.1 When Skeletal Musc *(I* Musc *Stretch My Mind)*

1. When does afferent sensory impulses increase from muscle?
   When change in muscle length (muscle spindle) or muscle tension (Golgi tendon)
2. When does the gamma efferent to musc become activated?
   Simultaneously with activation of alpha motor neurons
3. When does neurotransmitter at alpha motor neuron terminals get released?
   When calcium influx via *voltage-gated channels opened by arriving action potential*
4. When is action potential generated at skeletal muscle?
   When the end plate potential depolarizes the neighboring muscle membrane sufficiently to threshold firing
5. When is energy used during muscle contraction?
   When the myosin ATPase hydrolysis ATP
6. When is energy used during muscle relaxation?
   When calcium is re-sequestrated into the SR by calcium ATPase pump.
7. When does recruitment of motor units occur?
   When a greater number of alpha motor neurons in the spinal cord neuronal pool is activated.

8. When does muscle tetany occur during normal contraction?
   When increase tension of contraction is required and achieved by temporal summation of muscle contraction produced by increased freq of action potentials in alpha motor neurons.
9. When does muscle hypotonia occur?
   If the motor efferent is damaged or if the stretch reflex response is weak (hyporeflexia, since skeletal musc tone is assessed by the resistance to passive stretch performed by the neurologist)
10. When does blood flow to skeletal musc increase?
    When the active muscles produce vaso-dilator metabolites (K+ lactic) and local metabolic (<O_2, >CO_2) conditions also relax arteriolar smooth muscle. Adrenaline also vasodilates by binding to beta adrenergic receptors on the arteriole.
    There may also be cholinergic, vaso-dilator sympathetic fibers.
11. When is the musc able to extract more oxygen from the capillary blood?
    When the PO_2 gradient is increased as tissue PO_2 drops with higher O_2 usage
    When hemoglobin unloads more O_2; (Bohr CO_2/H effect) and warmer musc
12. When does muscle myoglobin release its store of oxygen?
    When the tissue PO_2 is much reduced since myGb-O_2 affinity is high and the P50 is much lower.
13. When is the muscle able to increase uptake of glucose from blood?
    When insulin is present for facilitated transport of glucose into cells.
    also insulin-independent, exercise-induced uptake of glucose mechanisms also contribute
14. When does bone release calcium from its exchangeable pool?
    When parathyroid hormone stimulate bone resorption as a response to hypocalcemia
15. When does skeletal muscle contraction increase, if any, when plasma catecholamine is high?
    Skeletal musc contraction is not dependent on hormones! Unlike important hormonal actions on cardiac and smooth muscles.

10.2 Muscle Sense

1. Obviously the brain initiates movement – ‘Dead men/women do not walk!’
2. Yet, for smooth, co-ordinated, accurate movement, the muscles sends information continuously back to the brain during muscle movement.
3. Two of these feedback sensory information from the moving muscles are *muscle length* and *muscle tension.* The muscle receptors that monitor length and tension are called *muscle spindles* and *Golgi tendon organs* respectively.
4. To illustrate muscle spindle receptor function in muscle length sensing, the simple reflex arc explains the mechanism.

5. The simple reflex arc is a monosynaptic circuit between the sensory nerve fibers from the muscle spindles (located inside a muscle 'motor unit') and the alpha motor neuron that causes contraction of the same group of muscle fibers (motor unit).
6. The muscle spindles (intrafusal fibers) are histologically parallel to the contracting (extra-fusal) muscle fibers.
7. Thus when a group of muscle fibers is stretched, the muscle spindles are also stretched and stimulated. The stretched muscle lengthens the muscle and this length change is conveyed by the stretched muscle spindle to the spinal cord via the sensory nerve fiber.
8. The muscle spindles actually send information about the muscle length changes and also the rate at which the muscle length is changing (associated with speed of muscle movement).
9. The alpha motor neuron in the spinal cord is activated by the action potential arriving in the sensory nerve fiber from the muscle spindles. Stimulation of the motor neuron always causes contraction of the group of muscle.
10. The other muscle sensors are located as the name indicates, in the muscle tendon that anchors the muscle to the bone. These tendon receptors senses muscle tension during contraction since the muscle are arranged in series with the tendon tension receptors.
11. Thus, the stronger the muscle contracts, the greater will be the tension monitored by the Golgi tendon organs and this sensory information is also sent to the spinal cord. The reflex response in the Golgi tendon reflex is to produce inhibition of the alpha motor neuron that innervates the same muscle group.
12. The Golgi tendon reflex is bi-synaptic and in situations of intense muscle contraction serves as a protective inhibitory reflex to prevent potential muscle tear and injury.
13. Keep in mind that muscle length information from the muscle spindles and muscle tension feedback from the tendon receptors during muscle movement are conveyed concurrently beyond the spinal cord level of the reflex to the higher centers in the brain.
14. The muscle spindle receptors have also their own, unique motor neuron innervations (gamma motor) that contracts the intra-fusal fibers. The function of the gamma motor activity is to maintain the sensitivity of the muscle spindle function during muscle movement.
15. When muscle movement is initiated from the brain, there is always co-activation of alpha and gamma motor neurons innervations the same group of muscles. ChengHMuscle011

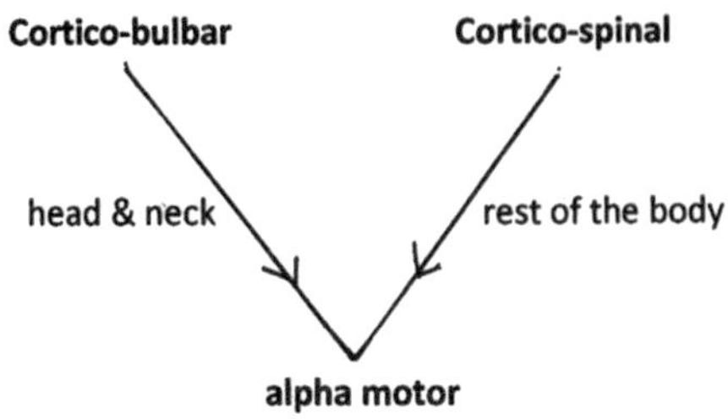

The alpha motor neurons are the final common pathway to the skeletal muscles. Descending cortico-spinal tracts arrive at the efferent spinal neurons that supply the corresponding muscles. The muscles of the neck and head region have their alpha motor neurons in the brain stem.

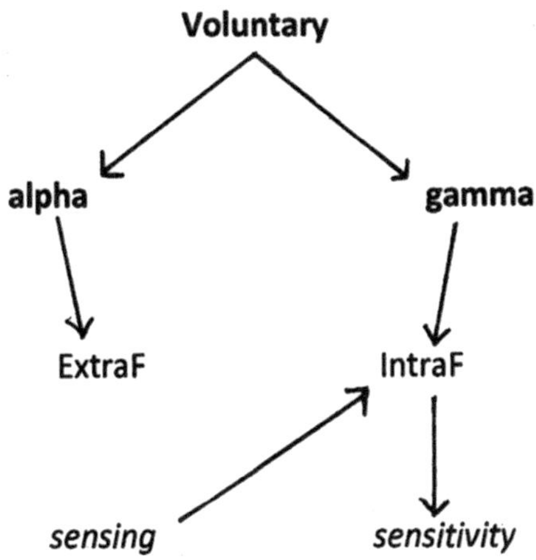

All intentional, voluntary actions are initiated in the brain. There is a concurrent co-activation of the alpha and gamma motor neurons to the skeletal muscles chosen to move. The simultaneous gamma coactivation with the alpha motor neurons ensure that sensory feedback from muscle movements are continuously sent to the cerebellum for fine-tuning and executing muscle motor activity in conformity with the cerebral intention.

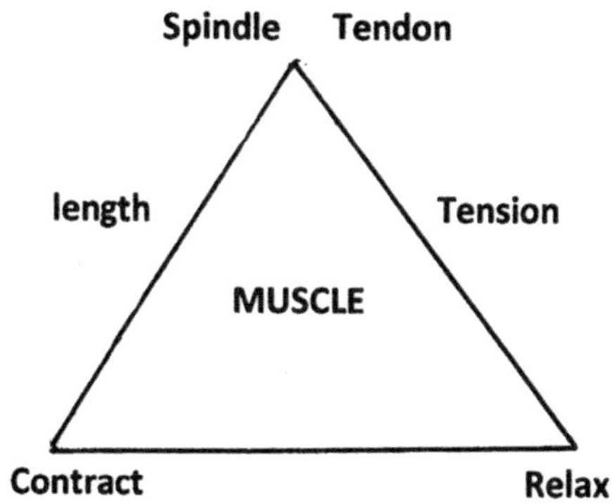

The two proprioceptors, muscle spindle and Golgi tendon provide sensory information to the brain of changes in muscle length and muscle tension during skeletal motor activity. The response of the monosynaptic muscle spindle reflex is contraction while the bi-synaptic tendon receptor reflex is inhibition of muscle contraction. Maintenance of posture and muscle movements both involve sensory feedback from these proprioceptors.

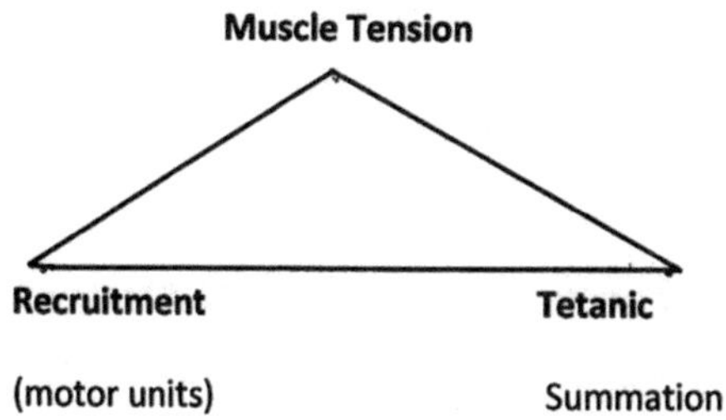

The grading of skeletal muscle tension during increasing strength of contraction is effected in two ways. There is motor unit recruitment as more efferent nerve fibers of the alpha motor neuron pool are activated. When all the motor units are recruited, increased frequency of action potentials can in addition produce tetanic summation to increase further the total muscle tension.

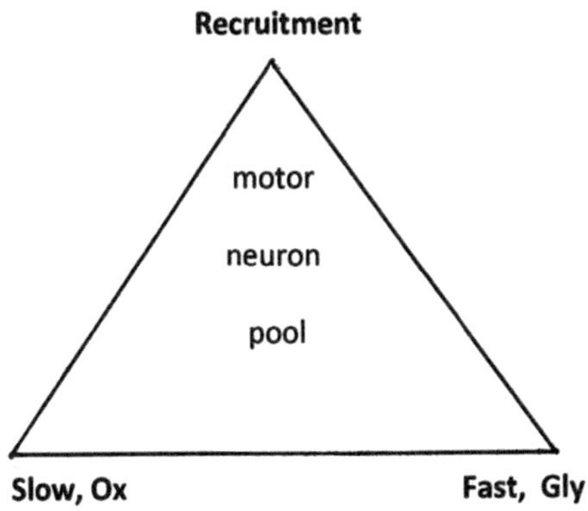

Recruitment of individual neurons in the spinal motor neuron pool is size dependent (size principle). Neurons with smaller cell bodies are activated to fire action potentials earlier. Recruitment of skeletal muscle types are also related to the size principle. The slow, oxidative muscle type or motor unit (each unit having only one muscle type), is recruited earlier than the fast glycolytic muscle fibers which are innervated generally by larger alpha motor neurons.

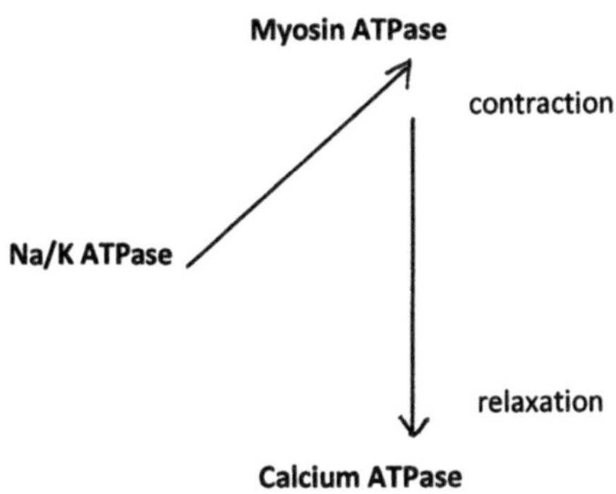

ATP provides the universal cellular energy for skeletal muscle function. Resting muscle maintains transmembrane ionic gradients by the NA/K ATPase. Actin-myosin sliding filamental contractions use myosin ATPase. A sarcoplasmic Ca-ATPase is needed to lower cytoplasmic calcium during skeletal muscle relaxation.

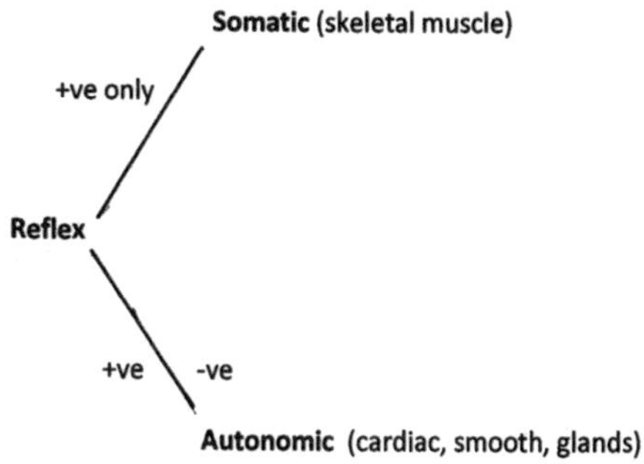

The word ‘reflex’ is a loose term that can apply to both voluntary and involuntary physiologic events. In skeletal muscle, reflex refers to the involvement of muscle spindle or tendon receptor reflexes in monosynaptic or polysynaptic motor responses. In autonomic reflexes, besides smooth muscles, cardiac muscle and both endocrine and exocrine glands are targets in the respective reflex action loops.

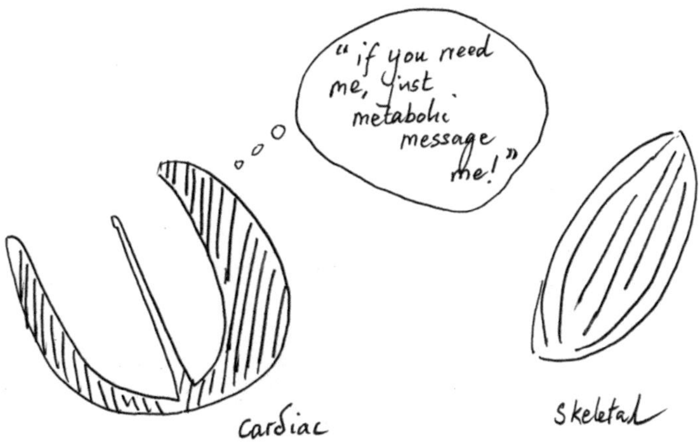

The cardiac muscle pump supplies increased blood flow as the metabolic demands of skeletal muscles increase during physical activity. How does the cardiac output keep in pace with the need for more perfusion to the skeletal muscles? The metabolite vasodilators autoregulate to increase the blood flow. Vasodilation involves the third conversant in this muscle talk. Its smooth talk! (vascular smooth muscle). Since the total peripheral resistance is lowered with the vasodilation, venous return is greater and that Starling-like, powers a higher stroke volume and cardiac output.

Part II
Endocrinology

Can you see the thyroids, ovaries, pancreas, testis and kidneys hanging out together?

Introduction: Hormones and Homeostatic Integration

Messages are constantly circulating in our blood. The bloody messages or endocrine hormones do not randomly bind to all cells in their pathway. The bioactive hormones have to exit the vascular space into the interstitial fluid to bind to its specific hormone receptors on the target cells. Target or responsive cells are those that have welcomed these messages. The messages received could be positive or negative but the receptive, hormone-submissive cells are activated to react accordingly.

Secretion of some endocrine hormones are directly influenced by its specific stimulus. There is like a 'push-pull' endocrine tango dance between the stimulus and the secretory response of the endocrine cells. In the current concern about 'metabolic syndrome' the relationship between blood glucose and the pancreatic beta cells is an example of this direct homeostatic endocrine interactions.

We can also think of the regulation of the two cations, calcium and potassium in ECF. Respectively, sensors for calcium and potassium are localized on the cell membrane of the endocrine cells that secrete parathyroid hormone and aldosterone.

In a more hierarchical control mechanisms, secretion of several hormones are orchestrated by the hypothalamus that conducts the pituitary glands to either secrete more or less of a particular hormone. The anterior pituitary secretes trophic or stimulating hormones e.g. thyroid stimulating hormone and prolactin. The posterior pituitary releases the two neuroendocrines, vasopressin and oxytocin, that are made by hypothalamic neurons.

Energy balance, Electrolyte balance and Body fluid balance are three aspects of homeostasis that require hormonal regulation. Some hormones like insulin and glucagon are anabolic and the balance is provided by catabolic hormones including glucagon, cortisol.

The desire to eat or appetite arises from activity of certain hypothalamic neurons which are in turn acted upon by orexigenic or anorexigenic mediators. During food consumption, hormones participate in the step wise digestive process as food is moved along the GI tract, e.g. gastrin, secretin, cholecystokinin.

The consistent service of the cardiovascular functions is supported by hormones. Physiologic concentrations of thyroid hormones contribute to a normal myocardial contractility and cardiac output. Optimal vascular responsiveness to enable appropriate resistance changes to blood flow is the effect of glucocorticoids from the adrenal glands.

ECF/blood volume is regulated by an interplay between anti-natriuretic and natriuretic hormones. The former endocrine actors are the RAS company led by renin.

Specific hormones make the male and female. The hypothalamus in the brain communicates hormonally with the testis and the ovaries. The female has the potential to secrete more hormones in her lifetime and this endocrine spectrum is increased by pregnancy-related hormones.

Chapter 6
Hypothalamus-Pituitary Hormonal Partners

The hypothalamus is like the watchman in a lighthouse, looking across the physiologic sea and surveying the conditions of the endocrine landscape. The homeostatic health and safety of the body requires the integrity of somatic/autonomic neural communication as well as endocrine signals.

The hypothalamus, in collaboration with its deputy watchman, the pituitary gland govern the circulatory messages that are sent out to peripheral endocrine glands to produce a wide spectrum of hormones. These hormones include those with principal action on metabolic rate (e.g. thyroxine), growth and repair of tissues (insulin-like growth factors), sex and reproduction (androgens, estrogens/progesterone), maintenance of normal cardio-vascular and renal functions (glucocorticoids).

The hypothalamus -pituitary hormonal partnership is integrated with the other roles of the hypothalamus in body fluid balance, energy homeostasis, sodium balance and activation of sympathetic nerve activity. Respectively, these regulations also include the secretions of hormones like vasopressin, orexigenic/anorexigenic factors, adrenal catecholamine and renal renin.

Mental states and moods also impinge on hypothalamus neurons and these can either enhance or suppress hypothalamic-pituitary activities. For example, in 'boarding school syndrome' the stress and disorientation of living away from home can result in 'missed periods', a temporary loss of the pituitary gonadotrophins that regulate ovarian/menstrual cycle. Adrenal cortisol is elevated as a physiologic response to stress.

1 How Does the Hypothalamus Regulate the Hormone Secretions from the Anterior Pituitary?

Answer The hypothalamus releases hypophysiotropic neuropeptides that control the secretion of hormones from the anterior pituitary. The hypothalamic neuropeptides either stimulate or inhibit their respective pituitary hormones.

H.M. Cheng (ed.), *Physiology Question-Based Learning*,
DOI 10.1007/978-981-10-0877-1_6

Concept The hypothalamus, called the 'master endocrine gland' orchestrates the hormonal rhythm and profile in response to environmental (e.g. light, temperature, offending sounds/sights) and afferent signals from the internal environment. Hypothalamic physiology includes the homeostasis of energy balance, fluid balance, body temperature, blood pressure and the sleep/wake cycles.

The hypothalamus is part of the diencephalon. At the floor of the 3rd ventricle, the two halves of the hypothalamus are rejoined to form a bridgelike area called the median eminence.

Some of the clusters of hypothalamic nuclei produce neurohormones. The term neurohormone indicates that these specialized neurons secrete neuropeptides which are released from their axon terminals in response to signals that depolarize the hypothalamic neurons.

The parvocelluar neurons in the hypothalamus have projections that terminate in the median eminence (also to the brainstem and spinal cord). The hypophyseal arteries supply arterial blood to the median eminence and the pituitary gland. The median eminence is also traversed by axons of the hypothalamo-hypophysial tract that terminate at the posterior pituitary.

The hypothalamic neuropeptides are 'eminently' secreted at the median eminence. There are long portal veins that drain the median eminence and these regulatory hypophysiotropic neuropeptides are transported from the primary capillary plexus to the secondary plexus that perfuses the anterior pituitary.

The hypothalamic neuropeptides include Corticotrophin Releasing Hormone (RH), Thyrotrophin RH, Gonadotrophin RH, Growth hormone RH, Growth hormone Inhibiting hormone (Somatostatin). The one exception to the family of hypophysiotropic neuropeptides is hypothalamic dopamine, a catecholamine that inhibits pituitary prolactin secretion.

In response, respectively, the anterior pituitary secretes a spectrum of peptide hormones including adrenocorticotropic hormone (ACTH, corticotropin), thyroid stimulating hormone (TSH, thyrotropin), follicle stimulating hormone and luteinizing hormone (FSH, LH, the gonadotropins), growth hormone and prolactin.

Together with any hormone secreted from the homeostatic neighbor, posterior pituitary (oxytocin, vasopressin), these hormonal messengers are transported in the venous blood draining the pituitary and enters the internal jugular vein to the systemic circulation.

2 How Does the Hypothalamus Affect Smooth Muscle Function via the Posterior Pituitary?

Answer The posterior pituitary secretes the two hypothalamic neuropeptides, oxytocin and vasopressin. Both neuro-hormones have vasoconstricting action on smooth muscles.

Concept The large magnocellular neurons that secrete neurohormones are clustered mainly in the paraventricular nuclei (PVN) and supraoptic nuclei (SON) of the hypothalamus. The two hormones, vasopressin (anti-diuretic hormone, ADH) and oxytocin are synthesized as preprohormones in the hypothalamic neuronal cell bodies.

These magnocellular neurons in the PVN and SON have long unmyelinated axons that end at the posterior pituitary. These axonal architecture are together called hypothalamo-hypophysial tract. Post-synthesis, the oxytocin or vasopressin (uncased in their vesicles with their neurophysins) are transported down their specific axons – is oxytocin or vasopressin synthesis localized to either the PVN or the SON?

Oxytocin has two major smooth muscle targets, in the lactating breast and the pregnant uterus. Parturition once initiated is sustained by the positive reflex arc, with afferent impulses arising from uterine mechanoreceptors, and arriving to stimulate hypothalamus-pituitary oxytocin release. Distention of the cervix just prior to parturition is also a stimulus for oxytocin.

After delivery, oxytocin agonists are used to help regress the uterus and also to reduce post-partum bleeding by vasoconstricting actions.

In the nursing breast, oxytocin is also positively involved in the milk ejection reflex, triggered by the mechano-receptors in the nipple that are stimulated when the baby feeds. The myoepithelial cells that line the alveoli and ducts in the mammary glands are contracted by oxytocin.

Apparently, it has been found that the neuropeptide oxytocin has an ultrashort, positive feedback action in stimulating its own release from the posterior pituitary. So remember, that the posterior pituitary secretes a 'positive' acting oxytocin.

While oxytocin satisfies the thirst of the baby, vasopressin participates in thirst and water balance. Both circulating vasopressin and oxytocin are nanopeptides. During negative water balance, changes in blood volume are sensed by both arterial baroreceptors and cardiac/pulmonary volume receptors. Afferent signals will activate the thirst neurons in the hypothalamus as well as increase the vasopressin secretion.

Besides conserving water in the kidneys, vasopressin acts on smooth muscle arterioles to add to the compensatory effect of increasing the total peripheral resistance. In the kidneys, the renal arteriolar constriction by vasopressin also slows vasa recta flow to the juxtamedullary nephrons for more effective urine concentrating ability during hypovolemia.

When we bring in the neighbouring anterior pituitary, the students might recall that the glucocorticoid cortisol, secreted in response to pituitary ACTH has a permissive action on smooth muscle vaso-responsiveness. And do not forget that increased sympathetic discharge associated with hypothalamic functions also target the peripheral arterioles (not including the coronary and cerebral) to constrict them. During thermoregulation, the hypothalamus selectively activates sympathetic vasoconstriction of cutaneous vessels to reduce heat loss during hypothermia.

3 How Is the Hypothalamus Involve in Sodium Physiology?

Answer The hypothalamus is the location of osmoreceptors and the neuroendocrine peptide vasopressin which together govern ECF sodium concentration/osmolarity. Sympathetic neural activation involving the hypothalamus participate in total body sodium or sodium balance.

Concept It seems at first thought strange to say that our brains are involved in salt (NaCl) regulation. The hypothalamus is essential in body fluid physiology. Water balance is maintained by the hypothalamus which have osmoreceptors that sense changes in water balance. Generally, changes in water balance, positive or negative are reflected in a lower or higher ECF sodium concentration. In turn, sodium concentration is the main determinant of the extracellular fluid osmolarity.

In other words, there are three synonymous terms that involve the same physiological control, namely water balance, osmolarity and sodium concentration. The thirst neurons that stimulate intake of water are also in the hypothalamus.

The operation of the hypothalamic osmoreceptors are linked to secretion of the water conservating hormone, vasopressin that is synthesized by hypothalamic neurons.

In addition to ECF osmolarity changes, isotonic volume changes also produce afferent sensory signals from baro- and volume receptors that both impinge onto the thirst center and the secretion of vasopressin.

Students commonly, loosely interchange the term sodium control physiology without specifying sodium Concentration or Total body sodium. A good illustration to distinguish the difference is ingestion of a large volume of water in a short time. The person will have her ECF hypotonically expanded. Although there is no change in her sodium balance, there is hyponatremia in the ECF.

In hypovolemic stress, the hypothalamus as the ‘head ganglion’ activates a general sympathetic discharge. The increased effector sympathetic activity includes the renal sympathetic neural arm (RSN). The action of RSN reduces urinary sodium excretion, conserves sodium in order to restore ECF and blood volume. Total body sodium is the main factor that determines ECF volume.

There are now descriptions of central sodium sensors that are likely integrated into the sympathetic control of sodium homeostasis. Components of the renin-angiotensin system e.g. angiotensin II have central actions.

The hypothalamus is thus ‘so dium’ important in preserving body fluid and ensuring an adequate intravascular blood volume.

4 How Is the Hypothalamus Involve in the Neuro-Hormonal Regulation of Cardiovascular Function?

Answer Cardiovascular function includes the cardiac pump and the vascular channels. The hormones from the adrenal glands, both and adrenaline and cortisol have effects on the cardiac muscle and blood vessels. Activation of sympathetic discharge and hypothalamo-pituitary hormonal axis to the adrenal glands secrete adrenaline and cortisol respectively. Renal sympathetic nerve stimulates secretion of renin which then generates the vasoconstrictor angiotensin II.

Concept Normal cardiac function requires a physiological level of cortisol. The responsiveness of the resistance vessels, the arterioles to vasoactive agents is also primed or 'permitted 'by the presence of physiologic plasma concentration of cortisol.

The hypothalamus secretes corticotrophin releasing hormone (CRH) which provokes the anterior pituitary to stimulate the adrenal cortex hormonally by ACTH action.

Normal cortisol is also required for a normal glomerular filtration rate (GFR). The relationship between GFR and cardiovascular (CVS) function is hinged on the control of ECF/blood volume by the kidneys.

The adrenal medulla secretes adrenaline when its cholinergic sympathetic innervations is stimulated. Eighty percent of the catecholamines that are released into the circulation is adrenaline and the rest, noradrenaline. Adrenaline and noradrenaline (NA) binds to beta receptors in the heart to produce chronotropic and inotropic, contractility effects.

Within the adrenal glands, cortisol upregulates the enzyme needed for conversion of noradrenaline to adrenaline.

The cardiac sympathetic nerve releases noradrenaline to give the tachycardic and increased ejection fraction. On blood vessels, most of the arterioles are served by mono autonomic sympathetic adrenergic vasoconstrictor fibers, its neurotransmitter, noradrenaline binds to alpha receptors.

In certain organs like the skeletal muscles, adrenaline acts on vascular beta receptors to vasodilate and increase blood perfusion. This occurs during exercise, when the sympatho-adrenomedullary unit is stimulated.

The hypothalamus also controls the secretion of its neurohormones, vasopressin and oxytocin. Vasopressin is released from the posterior pituitary when afferent signals, including hypovolemia are received by the hypothalamus. Vasopressin as its name implies is also a vasoconstrictor and it increases total peripheral resistance during hemorrhage.

Uterine post-partum bleeding is lessened by action of oxytocin on vascular smooth muscles.

The student should not neglect the secretion of renin by renal sympathetic nerve. Renin acts within the circulation to initiate the enzymatic generation of angiotensin II (AII) from angiotensinogen, a plasma protein precursor. AII is a potent vasoconstrictor and also has actions on cardiac tissues. The peptide AII also enters the brain and stimulates vasopressin secretion.

5 How Are the Hypothalamus and the Autonomic Sympathetic System Together Involve in Normal Physiology?

Answer The hypothalamus, together with the sympathetic nerves have important roles in cardiovascular, metabolic functions, thermoregulation and produce the physical effects of emotions.

Concept The autonomic system is quite often taught and learnt as a separate topic rather independently from any higher neuronal center input. This may not be the most ideal approach for a more realistic conceptual learning.

The hypothalamus has been described by Sherrington as 'the head ganglion of the autonomic system'. Activity of the hypothalamic neurons are certainly also expressed in many situations by changes in effector sympathetic activity. Here are some examples.

The hypothalamus is part of the limbic system that plays a role in a variety of diverse emotions. We know from common experience that excitement and anger are both associated with sympathetic -stimulated tachycardia.

Interestingly sexual excitement, besides triggering an increased sympathetic discharge also results in a specific parasympathetic vasodilatory effect on blood flow to the penis. Erection is produced. Other vascular arterioles are mostly regulated by varying the degree of sympathetic vasoconstrictor nerve fiber alone. This unique exception in parasympathetic vasodilation during male sexual activity explains why when we get angry we don't get a sympathetic induced erection!

Chronic lifestyle-related stress is associated with gastric/duodenal ulcer development. Gastric acid is increased by parasympathetic action on parietal HCl-secreting, gastrin and histamine-secreting cells. So although we think in the main about increased autonomic sympathetic reactions in 'fight or flight' scenarios, ulcerogenesis appears to be stress-related parasympathetic effect (alternatively, stress-induced breakdown of the gastric mucosal barrier).

The hypothalamus is also a central integrating place for body temperature control. Both peripheral cutaneous and hypothalamic thermoreceptors provide inputs. One major effector that compensates during thermoregulation is sympathetic nerve activity. Cutaneous vasodilatation to promote heat loss takes place when the sympathetic noradrenergic fibers are decreased. In contrast sweating that loses heat is stimulated by increased cholinergic sympathetic activity.

The hypothalamus also regulates energy balance. There are hypothalamic neurons that either stimulate (orexigenic) or inhibits (anorexigenic) food intake. Glucose sensing in the hypothalamus is liked to sympathetic nerve activity. Sympathetic activity participates in energy balance via its effect e.g. on hepatic glucose output and differential actions on insulin/glucagon secretion. In brown adipose tissues, there is extensive sympathetic innervations to both the fat cells as well as the blood vessels (Ganong).

6 Does the Hypothalamo-Pituitary Axis Govern Most Endocrine Hormone Secretions?

Answer The hypothalamus and the anterior pituitary regulates a variety of endocrine glands including, in the thyroid, liver, adrenal cortex, ovaries and the testis. The secretion of insulin and parathyroid hormone is not under the hypothalamo-pituitary control.

Concept The hypothalamus orchestrates via the anterior pituitary the appropriate rate of secretions of several hormones. Pituitary growth hormone (GH) acts on the liver to produce growth factors that have insulin-like anabolic effects (IGF-1). The GH is secreted in response to the net effect of two hypothalamic hormones, one stimulatory and an inhibitory (somatostatin) releasing factor.

The synthesis and secretion of sex steroid hormones from the ovary and the testis are stimulated by a dual combination of pituitary gonadotrophins, follicle stimulating hormone (FSH) and luiteinizing hormone (LH). FSH and LH as the name indicate are related to structures in the female ovarian/menstrual cycle.

Hormones tend to keep their historic names as their initial functions are discovered. Both FSH and LH have essential roles in males, during spermatogenesis and testosterone production respectively. A single gonadotrophin-releasing hormone (GnRH) from the hypothalamus stimulates FSH and LH.

The thyroid hormones T4, T3 have their specific pituitary thyroid stimulating hormone (TSH; thyrotrophin) which is in turn modulated by hypothalamic TRH (thyrotrophin -releasing hormone).

Pancreatic insulin that controls blood glucose and also potassium levels are not under the purview of the hypothamo-pituitary hormonal hierarchy. However, since the hypothalamus is associated with sympathetic activation, the suppression of insulin secretion by sympathetic activity spreads the hypothalamic influence also on pancreatic endocrine secretion.

Such is also the relationship between the hypothalamus and the adrenal medulla which secretes catecholamines. The sole neural input for the secretion of adrenaline is sympathetic cholinergic nerve.

The adrenal corticosteroid hormone, aldosterone is a mineralocorticoid that controls ECF potassium concentration and total body sodium. The pituitary adrenocorticotrophic hormone (ACTH) has a minor control on aldosterone secretion. ACTH primarily regulates the glucocorticoid, mainly cortisol that is released from the adrenal glands.

Since cortisol upregulates the enzyme activity that converts noradrenaline to adrenaline, this is a side hormonal pathway that links the hypothalamus-pituitary functions with the catecholamine secretion.

It appears that hormones that regulate electrolytes tend not to be part of the hypothalamo-pituitary axis of hormonal good. The parathyroid hormone that is a main player in ECF calcium homeostasis is secreted in direct response to hypocalcemia. Membrane calcium sensors on the endocrine cells in the parathyroid glands monitor the ECF calcium level. Thus there is no 'one hypothalamo-pituitary feedback loop' that governs all hormone secretions.

For sodium concentration, the Posterior pituitary-hypothalamus functional unit regulates by monitoring the ECF sodium concentration and responding accordingly. ECF osmolarity changes in parallel with sodium concentration. Hypothalamic osmoreceptor activity are linked to the secretion of vasopressin (anti-diuretic hormone), the hypothalamic neuro-peptide hormone, from the posterior pituitary.

7 In What Situations Is the Thyroid Gland Enlarged During Hyperthyroidism?

Answer Hypertrophy of the thyroid glands (goiter) occur when it is overstimulated by pituitary thyroid stimulating hormone (TSH) or thyroid stimulating immunoglobulin (TSI) that activates the follicle TSH receptors.

Concept TSH not only increases thyroid hormone synthesis and secretion but it also promotes growth of the endocrine glands. Excessive TSH secretion either from a hypothalamic or anterior pituitary abnormality would lead to a goiter and increased T3, T4 secretion.

The most common cause of hyperthyroidism is Graves' disease (called after Robert James Graves, an Irish doctor, nothing to do with the morbidity and the gravestone!). Graves' disease is due to an autoantibody. Unlike autoimmune diseases in which autoantibodies generally destroy the target tissues, the thyroid-stimulating immunoglobulin (TSI) binds to the TSH membrane receptors and triggers TSH-like cellular responses in the follicles.

The patient will develop goiter and has symptoms of hyperthyroidism. Unlike pituitary TSH, the TSI is not feedback regulated by the secreted thyroid hormones. In Graves' disease, the TSH level is reduced.

If the thyroid gland is hypersecreting due to a tumour, the primary cause leads to hyperthyroidism. Since the hyperactive gland is not due to overstimulation by a secondary rise in TSH, there would be no hypertrophy of the thyroid.

In addition, the TSH level is actually decreased due to negative feedback by the elevated thyroid hormones released from the gland. Physically, the gland can be enlarged due to the tumor growth but this is not a hypertrophied goiter that is due to either TSH or TSI trophic actions.

A unique presentation of Graves' disease not commonly seen in other categories of hyperthyroidism is protruding or bulging eyes (exophthalmos). This might be due to TSI action. There is water retention due to deposition of complex carbohydrates by fibroblasts behind the eye. Lymphocyte and fibroblast infiltration of the extra-ocular tissues contribute to the exophthalmos.

The hyperthyroid patient will have an elevated basal metabolic rate. The higher calorigenic effect will cause excessive perspiration and poor tolerance of heat. Although the patient will have increased appetite and eats more, body weight commonly falls due to the greater effect of increased metabolic rate. There is net catabolism of fat, carbohydrate and protein stores. The skeletal muscle proteins if affected, will result in muscle weakness.

The three general treatment methodologies for hyperthyroidism include use of antithyroid drugs to block specific biosynthetic sites, surgical removal of primary tumour and use of radioactive iodine, which is trapped by the iodide pump to then destroy hypersecreting glandular tissues.

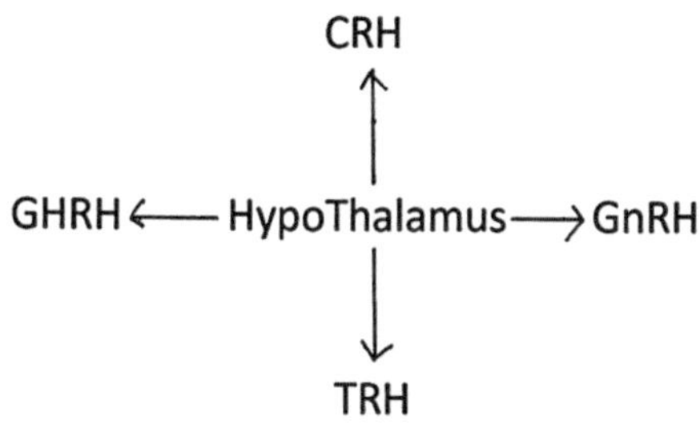

The four major hypothalamic releasing hormones that act on the anterior pituitary. All four are considered as neurohormones, made by neurons in the hypothalamus. The circadian rhythm of cortisol secretion is determined by the daily fluctuations of CRH. Growth hormone release, stimulated by GHRH also peaks during deep sleep. GnRH increases secretion of the two stimulating hormones in female and male reproduction, FSH and LH. TRH is a small tripeptide which releases thyroid stimulating hormone as well as prolactin.

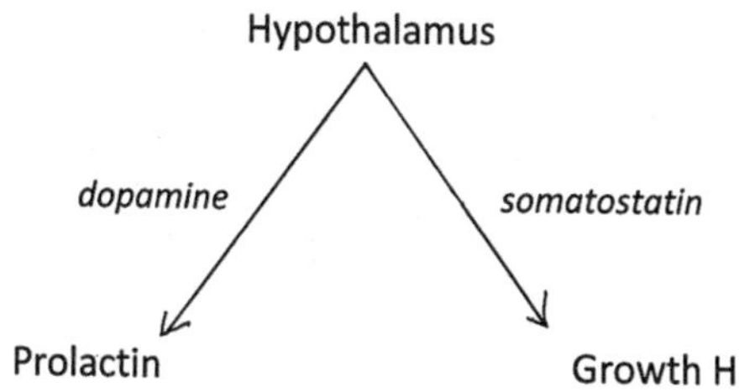

Two major hypothalamic control of pituitary endocrine secretions are inhibitory. One is hypothalamic dopamine regulator of prolactin for milk synthesis. For growth hormone (GH), its secretion from the anterior pituitary is under dual inhibitory/ stimulatory hypothalamic signals. Somatostatin inhibits GH secretion and somatostatin is itself stimulated by both GH and insulin-like growth factor (IGF 1 as part of homeostatic short and long feedback loops respectively. Does prolactin have a function in non-lactating females and in males?

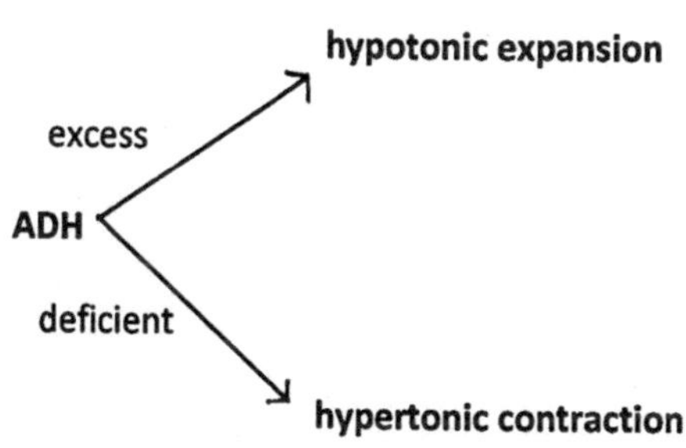

Anti-diuretic hormone (ADH) or vasopressin is a neuroendocrine secreted from the posterior pituitary. Uncontrolled secretion of ADH, usually due to a malignancy, leads to SIADH with water retention and both ECF and ICF volume expansion. Lack of ADH action causes increased urine volume and a hypertonic contraction. The urine is dilute (tasteless, insipid, hence diabetes insipidus). The patient in this case can replenish his fluid and restore her ECF osmolarity when thirst is stimulated at the hypothalamus.

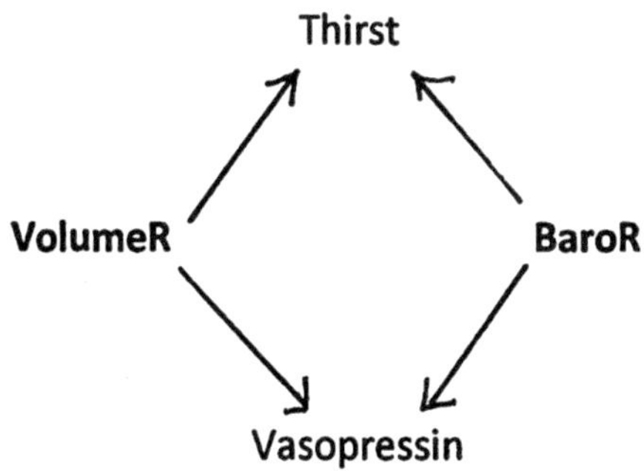

The recovery of fluid when the ECF becomes hyperosmotic or during isotonic hypovolemia is achieved by vascular volume sensors that include both the arterial baroreceptors and the cardiac/pulmonary volume receptors. A reduction in afferent impulses from the volume sensors produce two hypothalamic actions: increased thirst and also the release of vasopressin.

Chapter 7
Thyroid and Parathyroid Glands

The two lobes of the human thyroid are bridged by the thyroid isthmus tissue. Each thyroid follicle is encased by a single layer of functionally-polarized epithelial cells. The follicles are filled with a proteinaceous colloid that holds the store of the small amine thyroid hormones before they are enzymatically 'plucked' and released in to the circulation. Thyroid hormones have a general, priming, metabolic effect on virtually all cells in the body. Perhaps we could view physiologic levels of thyroid hormones as providing the cruise control of various body functions. Hypothyroidism leads to lethargy, slows mentation and reflex time is longer whereas hyperthyroidism increases cell metabolism, appetite is stimulated, anxiety and irritability can be heightened.

Humans have four small parathyroid glands closely associated anatomically with the thyroid glands. Careless thyroidectomy can remove the parathyroids, essential for calcium homeostasis. The endocrine Chief cells contain a prominent Golgi apparatus, endoplasmic reticulum and secretory granules. Human parathyroid hormone (PTH) in blood is a linear 84 amino acid polypeptide. The 3rd letter of the alphabet, C is for calcium. PTH has principal actions to elevate plasma calcium via its actions on 3 target organs, the bone, the kidneys and the intestines. The PTH activates vitamin D, a steroid hormone which then increases intestinal absorption of calcium.

The calcium lowering hormone calcitonin is produced by parafollicular cells of the thyroid glands. Calcitonin inhibits bone resorption and has been used in postmenopausal women to slow the progress of osteoporosis.

1 How Do Thyroid Hormones Affect the Cardiovascular System?

Answer Thyroid hormones, at physiologic levels are needed for normal cardiac output function. Thyroid hormones have sympathomimetic actions. The calorigenic effect of thyroid hormones is accompanied by promotion of heat loss via cutaneous vasodilation.

H.M. Cheng (ed.), *Physiology Question-Based Learning*,
DOI 10.1007/978-981-10-0877-1_7

Concept Physiologic concentration of thyroid hormones are required for cardiovascular function. Hypothyroid individuals have decreased cardiac output.

Some of the cardiac actions of thyroid hormones are associated with their enhancement of metabolic rate with increased oxygen consumption. The metabolic actions produce heat, a calorigenic effect of thyroid hormones.

The thyroid hormones are regulated by a hypothalamo-pituitary endocrine control axis.

Although a primary action of thyroid hormones is elevation of metabolic rate and generation of heat, an acute rise in thyroid stimulating hormone (TSH) from the anterior pituitary is not observed when an adult is exposed to cold. However, newborns respond to cold with an increase in TSH that is stimulated by an increase in hypothalamic thyrotrophin releasing hormone (TRH). TRH binds to membrane receptors and acts via the inositol triphosphate/calcium/diacylglycerol second messenger pathway.

Thyroid hormones have sympathomimetic actions. This is brought about by an upregulation of adrenergic receptors on target tissues. Thyroid hormones bind to nuclear receptors and activate transcription of proteins that include receptor proteins and cellular metabolic enzymes. The increased cardiac output can be explained by tachycardia and increased myocardial contractility (bigger stroke volume) due to more adrenergic receptors binding in both the cardiac sino-atrial pacemaker cells and the ventricle muscles respectively.

The longer term effect of inducing a change in the cardiac myosin ATPase isoform also contributes to a greater ejection fraction.

The use of beta-blockers to alleviate symptoms during an excessive thyroid hormone 'crisis' 'or 'storm' is based on this synergistic effect of thyroid hormones on sympatho-adrenal functions.

A prolonged period of higher cardiac output strains the heart with a greater cardiac workload. High cardiac output failure can occur in hyperthyroidism.

The metabolic/calorigenic effects of thyroid hormones increase body temperature and set in action heat loss mechanisms. Vasodilation of blood vessels supplying the skin promotes heat loss from the body's surface. The bigger cardiac output with increased thyroid hormones is also accounted for by the reduced total peripheral resistance (tpr), leading to a greater venous return.

Remember to 'Thy a ribbonoid' around your heart!

2 What Membrane Protein Functions at the Basolateral Side of the Thyroid Follicle Are Essential for the Initial Steps of Thyroid Hormone Synthesis?

Answer Iodide enters the follicle cells from the capillary blood by secondary active transport. Tyrosine is also transported into the follicle cells by sodium-amino acid co-transporter.

Concept The thyroid hormones are unique in their production in requiring an essential dietary element, iodine. Both triiodothyronie (T3) and tetraiodothyronine (T4, thyroxine), as the name indicate, are iodated tyrosine hormones.

At the baso-loateral membrane of the thyroid follicle, the non-essential amino acid tyrosine and the iodide are transported into the follicle epithelial cells. The apical membrane of the follicle faces an inland, extracellular follicular lumen where thyroglobulin is stored in the colloid. The final steps of thyroid hormone synthesis occurs on the large thyroglobulin protein molecule itself when tyrosine residues of the thyroglobulin becomes iodinated.

Circulating tyrosine likely enters follicle cells by a sodium-linked tyrosine symporter as occurs in other epithelial cells that transport amino acids e.g. at the nephron and the intestinal enterocyte. The ubiquitous sodium/potassium ATPase powers the sodium-coupled amino acid cotransporter.

The iodide also enters at the basolateral membrane by secondary active transport. There is a sodium-iodide 'pump' that serves to concentrate iodide in the thyroid follicle cells. The transmembrane sodium gradient sustained by the Na/K ATPase energizes the influxes of the two ingredients for thyroid hormone synthesis, both the iodide and the tyrosine.

The megaprotein thyroglobulin is made by the endoplasmic reticulum/Golgi complex in the thyroid follicular cells. The tyrosine-rich thyroglobulin is then exported into the colloidal lumen by exocytosis.

In contrast, note that in endocrine cells that secrete peptide hormones, exocytosis of vesicles is the process that releases the peptide hormones into the circulation. For triidodothyronine and tetraiodothyroninem, their secretion into the circulation upon stimulation is via simple diffusion from the follicle cells since both thyroid hormones are lipid soluble.

The thyroid follicle cells are stimulated to secrete when thyroid stimulating hormone (TSH) from the anterior pituitary is increased. The membrane receptors for TSH binding are also located on the baso-lateral interface of the follicle cells.

The thyroid gland has a bow-tie shape, with the two lobes of endocrine tissues connected by an isthmus. It is coincidently situated also at the position of a bow tie, lying just below the larynx, over the trachea. Applause the Tieroid gland that secretes essential amine hormones (T3, T4) and he will take a bow!

3 At the Luminal Surface of the Thyroid Follicle, What Membrane Events Are Involved in the Synthesis of Thyroid Hormones

Answer Iodide trapped inside the follicle cell are oxidized to its free radical by the luminal membrane enzyme thyroperoxidase (TPO). Iodination of tyrosine groups on thyroglobulin and coupling reactions to form T3 and T4 Are also catalyzed by TPO. Endocytosis of bits of hormone-rich thyroglobulin colloid via engulfment by the luminal membrane is stimulated when secretion of T3/T4 is needed.

Concept The sodium-iodide secondary active symporter at the baso-lateral membrane concentrates and traps iodide inside the thyroid follicle cell. The iodide anion then diffuses down its electrical and concentration gradients towards the luminal side of the follicle cell.

At the luminal surface, a membrane enzyme, thyroperoxidase (TPO) rapidly oxidizes the iodide to a reactive free radical.

The membrane TPO then iodinates the tyrosine residues of thyroglobulin. The phenolic ring of a tyrosine is either iodinated once (monoiodotyrosine, MIT) or twice (diiodotyrosine, DIT).

There is then a coupling reaction, likely also mediated by TPO between MIT and DIT on the thyroglobulin molecule (couple married on the thyrotanic!). Coupling of two DIT yields thyroxine (tetraiodothyronine). A MIT and DIT couples together to give triiodothyronine (T3).

Thyroid hormones are unusual in being stored in the colloid until needed. The reservoir pool of available T3/T4 is sufficient for several weeks even if the diet is iodine-deficient.

When thyroid gland is stimulated by TSH, almost all the biosynthetic processes in the follicular cell are activated. At the luminal surface, endocytosis of a blob of thyroglobulin (Tg) colloid takes place.

Inside the follicle cell, lysosomes approach the engulfed thyrogobulin and proteolysis occurs to release the thyroid hormones (T3, T4), MIT and DIT. The latter two iodotyrosines are inactive but can be recycled for further synthesis of thyroid hormones. Interestingly, an intracellular iodinase removes iodide from MIT, DIT but does not physio-logically touch T3 and T4.

The lipophilic T3 And T4 diffuse at the basolateral membrane into the interstitial fluid and then into the capillary blood, where they circulate, mostly bound (>99 %) to plasma proteins. The main protein carrier is TBG (thyroxine-binding globulin).

TSH is also a trophic hormone. Besides stimulating the synthesis and secretion of thyroid hormones, it also increases e.g. protein synthesis, DNA replication, cell division of the thyroid follicles. Thus excess secretion of TSH leads to hypertrophy of the glands (goiter). This enlargement of the gland takes place also in the absence of thyroid hormone secretion as seen in iodine-deficiency goiter. The loss of T3/T4 hormonal feedback results in the pituitary TSH hypersecretion.

4 How Does the Regulation of Insulin Secretion Illustrate the Multiple Types of Stimuli that Modulate Endocrine Cell Secretion

Answer The variety of inputs that directly act on endocrine glands include autonomic nerves, other paracrine and endocrine hormones, ions/nutrients and sensory receptors.

Concept Hormone secretion occurs by exocytosis for peptide hormones and catecholamines. For steroid and thyroid hormones, secretion is by diffusion from the endocrine cells into the interstitial space and hence into the capillary blood.

There can be four categories of stimulus that can influence an endocrine cell secretion. For any endocrine hormone, more than one stimulus type commonly regulates the hormonal secretion.

Let's consider the hypoglycemic hormone insulin. Incidentally, insulin is uniquely the only hypoglycemic hormone. All the other hormones that have a role in glucose homeostasis (glucagon, adrenaline, cortisol, growth hormone) are hyperglycemic, oppose insulin action and are collectively named 'counter-regulatory' glucose controls.

Within the islet of Langerhans, there is paracrine fine tuning of insulin secretion from the beta pancreatic cells. Locally, in the pancreatic endocrine cell community, somatostatin inhibits and glucagon stimulates insulin secretion. A class of gastrointestinal hormones termed 'incretins' are released by luminal glucose. Before any absorbed glucose appear in the postprandial blood, incretins as its name indicate, stimulates a preparatory feed-forward secretion of insulin. The two major incretins are GIP (glucose-dependent insulinotropic peptide; other functional name is gastric inhibitory peptide) and GLP-1 (glucagon-like peptide 1).

Reduction in incretins, after a meal is absorbed, also allows insulin to decrease even when blood glucose is still higher than in fasting levels. This could be seen as a physiologic design to prevent insulin-induced hypoglycemia.

The parasympathetic nerve stimulates insulin secretion. Conversely, sympathetic stimulation and adrenal catecholamines both act on the same adrenergic receptors on pancreatic beta cells to inhibit insulin secretion.

Insulin is secreted by post prandial hyperglycemia. Glucose is 'sensed' indirectly via ATP production and ATP-sensitive potassium channels which leads to depolarized pancreatic beta cells. Insulin is then exocytosed, triggered by the calcium influx via voltage gated channels. Amino acids are the other main nutrient stimuli for secretion of the anabolic insulin. Does fatty acids, most transported as lipoproteins, also stimulate insulin that has a lipogenic action?

Besides hormonal, neural and nutrient stimuli, insulin is also hypokalemic and this serves to prevent excessive postprandial hyperkalemia. Insulin stimulates activity of the ubiquitous membrane Na/K ATPase to increase cellular uptake of potassium.

For some hormones, sensory receptors are linked to their secretion. Here we can think of vasopressin and its rate of secretion varying with afferent impulses from the volume/pressure vascular receptors.

Mechano-sensing by endocrine cells themselves also exist in physiology. The secretion of cardiac natriuretic hormones by atrial/ventricular volume distension is an example. The renal arterial perfusion pressure is monitored by 'intra-renal baroreceptors' at the afferent arteriolar cells. The hormone renin is secreted when hypovolemia/hypotension lowers the renal perfusion pressure.

5 How Is the Remodeling of Bones in Cathy's Body Accomplished?

Answer There is a continuous turnover of living bone tissues achieved by an interplay between matrix-making osteoblasts and matrix-resorbing osteoclasts in bone. Osteoblast secretes signals that can either increase or reduce osteoclast bone resorption activity.

Concept Ninety nine percent of the body's calcium is in bone. Our in vivo skeleton is a living tissue and the bone marrow produces the blood cells, including erythrocytes, leucocytes and platelets. The organic extracellular matrix of bone or osteoid is impregnated with hydroxyapatite crystals. These crystals are precipitates mainly of calcium phosphate which crystallize around collagen fibers in the osteoid. There is an inter-changeable reservoir pool of bone calcium that can be mobilized by resorption when hypocalcemia occurs. The hormone from the parathyroid glands is the key endocrine factor that maintains blood calcium.

There are three types of cells in bone that collaborate in bone remodeling and maintain bone mass or density. The osteoblasts secrete the osteoid matrix materials that are the base for calcium phosphate hydroxyapatite crystal formation. Osteoblasts become osteocytes, aged or retired osteoblasts, caged within the bony structures they have made around themselves.

The second cell type is osteoclast, which is multinucleated and acts on the organic matrix to resorb bone. The osteoclast attaches itself to bone and secretes hydrochloric acid (like the gastric parietal cells) and enzymes that solubilize the calcium phosphate crystals and matrix. The effect of osteoclast action is to release calcium into the circulation.

The balance in bone remodeling is thus between the see-saw, opposing actions of osteoblasts and osteoclasts. Osteoblasts are derived from bone stromal cells while the phagocytic osteoclasts differentiate from macrophages which come from the monocyte lineage.

Interestingly, the osteoblast secretes two major signals that determine the degree of its counterpart's (osteoclast) development and activity. A molecule called RANKLigand (RANKL) binds to bone tissue macrophages and stimulates differentiation into osteoclasts. RANKL also prolongs the lifespan of osteoclasts (delays apoptosis) to further the resorption of bone. You could say that RANKL from osteo-Blasts 'cRANKed up' osteoClastic activity.

The other signal from osteoblasts, osteoprotegerin (remember 'osteo integrity') has the net effect of promoting osteoblast bone building activity and increasing bone density. The action of osteoprotegerin (Opg) is to complex with RANKL and this then prevents RANKL from acting on macrophages to transform them into osteoclasts.

Cathy's hormone, oestrogen preserves bone and favours osteoblastic events. Indeed oestrogen has been found to increase transcription of Opg gene in osteoblasts. Oestrogen also speeds up the apoptosis of osteoclasts to strengthen Cathy's bone integrity.

6 How Does the Phosphaturic Action of PTH Hormone Contribute to Calcium Homeostasis?

Answer Parathyroid hormone (PTH) acts to resorb bone and increases intestinal absorption of calcium. To achieve a net increase in free ionized calcium in blood, the increased excretion of urinary phosphate by PTH is essential.

Concept Calcium hormeostasis is regulated by three hormones, namely PTH, vitamin D and calcitonin. There are also three target tissues involved in the hormonal regulation of ECF calcium; the kidneys, the bone and the intestines.

The calcium sensors that monitor ECf calcium are located on the endocrine cell membrane in the parathyroid glands. Hypocalcemia is detected and PTH is secreted. The net effect of PTH action is to raise plasma calcium to normal.

Circulating calcium is present in three forms, plasma protein bound, complexed with other anions and free ionized calcium. The bioactive calcium is the ionized species. The calcium sensors respond to the free calcium level.

When PTH is secreted during hypocalcemia, the hormone acts on the three target organs to normalize the free calcium concentration. PTH acts on bone to release calcium from calcium interchangeable skeletal stores. Since calcium phosphate is the main inorganic mineral in bone, both calcium and phosphate are released and increased in blood by the resorption of bone.

The intestinal absorption of calcium is indirectly mediated by PTH via the synthesis of active vitamin D in renal endocrine cells stimulated by PTH. Calcitriol or 1,25 dihydroxycholecalciferol, the active vitamin D increases active calcium absorption by the enterocyte epithelial cells. Intestinal absorption of dietary phosphate is also increased by calcitriol.

The actions of PTH on the bone and the intestines increase the blood calcium phosphate concentration. The free bioactive calcium, to be increased requires the additional action of PTH in the kidneys. At the nephrons, both non-protein bound calcium and phosphate are freely filtered. Filtered calcium is reabsorbed by the renal tubules under PTH action.

Conversely, PTH specifically inhibits the reabsorption of tubular fluid phosphate. The urinary excretion of phosphate is increased. PTH can be remembered as a PhosphaTuric Hormone. The tubular action of PTH is cyclic AMP mediated. Thus the overall, net action of PTH, secreted in response to hypocalcemia, on bone, intestines and kidneys is to increase the free ionized calcium concentration.

Tubular fluid phosphate, the student might recall is a urinary buffer involved in ECF pH balance. A side physiologic role of PTH is then an increase in the urinary phosphate buffer capacity to bind and excrete hydrogen ions.

7 What Two Hormones Produced by the Liver Are Involved Directly or Indirectly in Blood Cells Production?

Answer The liver produces thrombopoietin which stimulates platelet production. The hepatic hormone hepcidin negatively regulates intestinal iron absorption to ensure sufficient iron for hemoglobin synthesis.

Concept Almost all intestinal iron absorption occurs in the duodenum. Transporters at both the apical and basolateral membrane of the enterocyte epithelial cells participate in iron absorption. Dietary iron is mainly in the ferric form and the ferrous iron is the form that enters the enterocytes. A membrane reductase does the metallic job!

Trans -membrane movement of the iron cation into the enterocyte occurs via the divalent metal transporter (DMT). Iron then exits the enterocyte by a basolateral transporter named ferroportin. The enterocytes in iron deficient animals absorb more iron than normal and this is regulated by the hepatic hormone hepcidin. What operates as the sensors for iron stores in the body and for perhaps the erythropoietic status in the bone marrow?

Hepcidin functions as a negative regulator of intestinal iron absorption. When the requirement for iron is increased e.g. following blood loss, the circulating hepcidin level is low. Hepcidin action might negatively determine the number of active DMT and ferroportin at the enterocyte.

The other source of iron in dietary meats is iron complexed with heme. At the apical membrane, there is a heme transporter. In the cytosol, the heme oxidase releases the iron. The destiny of the iron can be down two pathways, depending on the body's iron status. Most of the iron is converted back to its ferric form and stored in the enterocytes, bound as ferritin. The daily slough of exfoliated enterocytes that end up as part of the excreted fecal material means you have iron output in your toilet!

In vivo, note that the heme in hemoglobin that is released from aged, expire red cells are metabolized in the liver to bilirubin.

Platelet production is also under hepatic control. The precursor cells that give rise to platelets are the megakaryoblasts. Differentiation of the megakaryoblasts changes them to megakaryocytes. Thrombopoietin appears to act to mature the megakaryocytes and to stimulate the shedding of fragments from this mega cell to become what would then circulate in the blood as thrombocytes.

To add on to the hormonal functions of the liver on blood erythrocytes and platelets, the liver is also the major site for synthesis of the coagulation proteins. The students will recall that platelets play an initial role in hemostasis in the formation of platelet plugs (Platelets Unplugged would make an entertaining, physiologic comic drama!).

Vitamin K -dependent hemostatic factors are made in the liver, and the absorption of the fat-soluble vitamin K requires the micellar transport within the intestinal lumen. This is achieved by the hepatic bile, released into the duodenum by gall bladder expulsion.

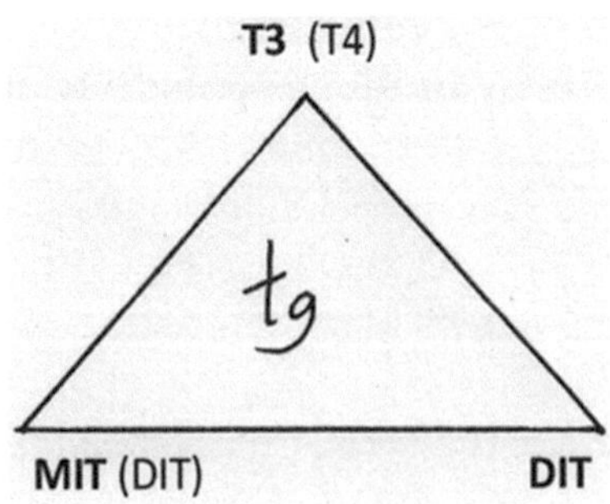

Biosynthesis of thyroid hormones is unique in requiring an essential dietary element, iodine. Equally fascinating is that the production of the hormones in the thyroid follicles occur at the colloidal stage, on the body of the 'globulin' (thyroglobulin, Tg). There are four different 'aminated' actors (iodinated, amine hormones) on thyroglobulin, holding one to four iodine molecules. Specific couplings among two of the pre-hormonal actors (MIT and DIT) generate the thyroxine (T4) and triiodothyronine (T3).

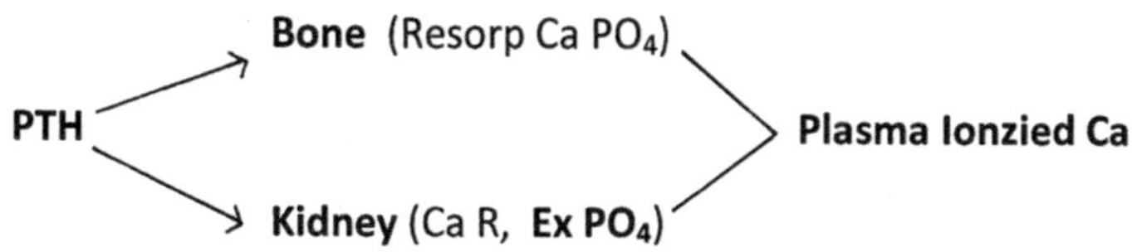

The net effect of the triple actions of parathyroid hormone on bone, intestine and kidney is to raise the free ionized calcium. PTH resorps and releases calcium and phosphate from skeletal bone. Its indirect effect via activated vitamin D increases both calcium and phosphate intestinal absorption. Specifically, PTH reabsorbs the filtered calcium but inhibits the filtered phosphate. The net effect of this phosphaturic hormone ('PTH') is to increase and normalize ECF free calcium, the bioactive, controlled calcium parameter.

What Parathyroid glands say to the intestinal Enterocyte. The intestinal absorption of calcium is increased by vitamin D. The active form of vitamin D is generated in the kidneys when parathyroid hormone catalyses the enzymatic hydroxylation.

T_4hyroid@Hot$_3$mail $^{\circ}$Com

This physiomail is a useful game for students to try to summarize some key concepts in Physiology. Here the Thyroxine (T4) is often converted by a de-iodinase in target cells to triiodothyronine (t3). The principal action of the thyroid hormones is to increase basal metabolic rate, producing heat that contributes to body temperature.

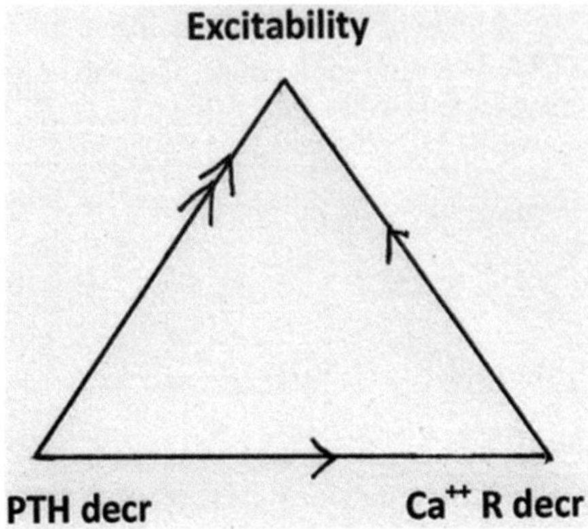

Parathyroid hormone (PTH) is the key polypeptide hormone that prevents hypocalcemia. ECF calcium is carefully controlled at a low level (~2.5 mmol/L). Hypocalcemia, seen in hypoparathyrodism results in increased responsiveness of excitable cells. Hypocalcemic tetany is a common symptom.

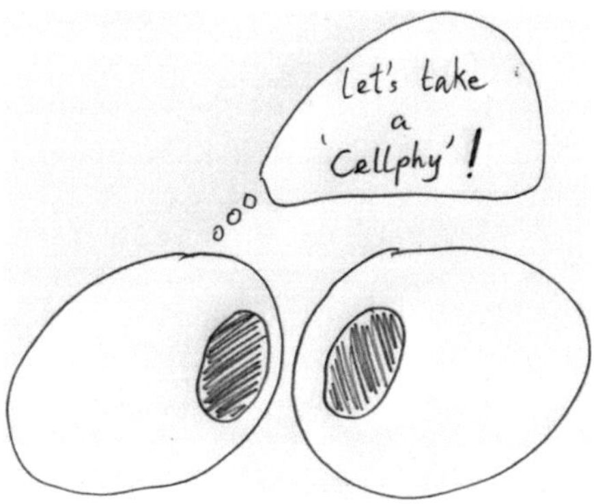

Selfie anyone? This cartoon depicts the cell-cell crosstalk by way of paracines, besides any extracellular matrix interactions.

Chapter 8
Adrenals and Pancreas

The adrenal gland is hormonally, two endocrine glands, the cortex and the medulla. Other than the male and female reproductive hormones, the adrenal cortex synthesizes and secrete steroid hormones that all have the cholesterol framework (good cholesterol!). The three classes of corticosteroids are androgen, the glucocorticoid and the mineralocorticoid. The latter is represented by aldosterone the key hormone that has a dual homeostatic role in potassium and sodium balance. Unlike the adrenal glucocorticoid and androgen, aldosterone is not under strict control by a hypothalamo-pituitary axis.

Cortisol, the dominant glucocorticoid is essential for life. In hypoglycemic stress, cortisol is secreted to produce metabolic effects on skeletal muscles and the liver to sustain blood glucose levels. The catecholamines from the adrenal medulla also have hyperglycemic actions and these amine hormones, made from the amino acid, tyrosine are secreted when the autonomic sympathetic nerve to the adrenal medullary cells are stimulated. Circulating adrenaline and neurotransmitted sympathetic noradrenaline are supplementary in their actions on cardiac, vascular and renal functions in arterial blood pressure regulation.

Indeed, the pancreatic insulin holds a unique homeostatic position as the only hypoglycemic hormone in blood glucose control. All other hormones involved in glucose homeostasis are counter-regulatory to insulin action, including glucagon, growth hormone. The final absorbable products from digestion of fats, proteins and carbohydrates by the enzymes from the exocrine pancreas all stimulate insulin release. Insulin is an anabolic hormone that promotes energy storage as adipose fats, protein synthesis and liver/muscle glycogenesis (insulin sounds like 'insurance'!).

Insulin from the pancreatic beta cells also acts in concert with adrenal aldosterone in potassium homeostasis. There is ongoing studies to elucidate a role of insulin in the kidneys that synergizes with aldosterone in renal handling for sodium homeostasis.

H.M. Cheng (ed.), *Physiology Question-Based Learning*,
DOI 10.1007/978-981-10-0877-1_8

1 How Are the Adrenal Hormones Involve in Arterial Blood Pressure Control?

Answer The adrenal medulla secretes sympathomimetic catecholamines that act on the heart and blood vessels. The glucocorticoid cortisol sensitizes the blood vessels for optimal vascular responsiveness. The mineralocorticoid aldosterone maintains sodium balance/blood volume to maintain an adequate cardiac output.

Concept Most of the catecholamines secreted from the adrenal medulla is adrenaline. Circulating adrenaline complements the action of cardiac sympathetic nerve by binding to the same beta adrenergic receptors on the myocardium and the pacemaker sino atrial cells.

The adrenal cortex secretes three categories of steroid hormones, the mineralocorticoids, glucocorticoids and the androgens. In blood pressure regulation, the determinants are the total peripheral resistance (TPR) and the ventricular cardiac output (CO). The major glucocorticoid, cortisol is needed at physiological concentration for its permissive action on arteriolar response to vaso-active agents. The requirement for this vaso-priming effect is seen in either excess or deficiency of cortisol. In Cushing's syndrome, a hyperresponsive blood vessel contributes to the hypertension. Conversely, in primary adrenal insufficiency (Addison's disease, where there is Minus of all the adrenocortical hormones!), hypotension is not only due to the volume contraction of aldosterone lack but also due to the hyporesponsive arterioles.

The major mineralocorticoid, aldosterone acts to ensure a normal sodium balance. Sodium balance is the main determinant of ECF/blood volume. The cardiac output determinant in the blood pressure relationship (BP = CO x TPR) is homeostatically under aldosterone's hormonal supervision.

Secretion of the adrenal aldosterone, cortisol and catecholamines are in turn modulated by their respective A factor. For aldosterone, it is Angiotensin II, a peptide product during activation of the renin-angiotensin system. For cortisol, the A factor is Adrenocorticotrophic hormone (ACTH), secreted from the anterior pituitary. The Autonomic cholinergic sympathetic nerve fibers innervate the adrenomedullary endocrine cells.

The adrenal glands is also called supra-renal glands. The close proximity reminds the students of the integrated function between the kidneys and the adrenals in regulation of arterial blood pressure. The kidneys are essential in long-term blood pressure mechanisms via its key role in maintain the sodium balance. The renal arterioles are also vascular targets in selective constriction by sympathetic nerve during compensatory increase in TPR. The cortisol hormonally 'permits' the vasoconstriction.

Does the 'permissive action 'of cortisol extends also to the vasodilatory effects of renal paracrines like prostaglandins or of the local natriuretic urodilatin that also reduces the renal arteriolar resistance?

The major adrenal androgen in both male and female is DHEA (dihydroepiandrosterone). DHEA is a much weaker androgen compared to testosterone. The latter testicular male hormone does stimulate erythropoiesis which accounts for the higher hematocrit in males.

2 What Is the Potential Secondary Problem of Having One Key Hormone, Aldosterone Controlling Two Major Electrolyte Balance?

Answer If there is a primary excess of aldosterone secretion in a patient, there will be concurrent positive sodium and negative potassium balance. However, in normal individuals, a negative sodium balance is resolved without a secondary hypokalemia.

Concept The adrenal corticosteroid hormone, aldosterone is the key player in the regulation of the two cations in the ECF, sodium and potassium. Sodium balance, linked to ECF volume and potassium concentration in the ECF are both maintained by aldosterone.

For potassium homeostasis, membrane sensors are located on the endocrine cells in the zona glomerulosa of the adrenal cortex. Hyperkalemia is directly sensed and this leads to secretion of aldosterone. Aldosterone then increases the active tubular secretion of potassium at the collecting ducts of the nephron.

For sodium balance, the trigger that stimulates aldosterone release is the circulating peptide, angiotensin II that is generated when the renin-angiotensin pathway is activated.

Aldosterone acts on the same principal cells of the collecting ducts to increase active sodium reabsorption and potassium secretion. At the baso-lateral membrane the common Na/K ATPase activity is heightened. At the luminal membrane of the principal cell, the permeability to sodium and potassium is increased by an increase in sodium and potassium channels respectively.

This dual hormonal action of aldosterone poses the obvious question: Will a primary sodium or potassium imbalance leads to compensatory aldosterone action that resolves the electrolyte disturbance but ends up with a secondary cation problem? This is called the 'aldosterone paradox'.

Physio-logically, this does not occur. A negative sodium balance is accompanied by hypovolemia, with a reduction in renal blood flow (RBF). The decreased RBF lowers the glomerular filtration rate and tubular fluid flow is consequently less. Since the luminal step of tubular epithelial secretion of potassium is by passive diffusion, a decreased tubular fluid flow will retard potassium secretion.

This is the reason why when hypovolemia activates the renin-aldosterone mechanisms, the increased tubular reabsorption of sodium is not followed by a parallel increase in potassium secretion and a resultant secondary hypokalemia.

Similarly, a primary hyperkalemia stimulates aldosterone secretion. The normalization of potassium balance is also not associated with a secondary sodium retention. Here there appears to be a direct action of hyperkalemia to inhibit proximal tubular transport of sodium (mechanism still to be elucidated) from the tubular fluid.

Pathophysiologically, a hypersecretion of aldosterone will give double Cationic trouble (Meow, Meow!). In Conn's syndrome, there is ECF volume expansion from sodium retention as well as hypokalemia from hyperkaliuria.

Conversely, in Addison's disease, a deficiency of aldosterone causes hyperkalemia and a hypotonic contraction of the ECF in the patient. The patient is hypotensive due to a reduction in blood volume.

3 What Is the Phenomenon Described as 'Aldosterone Escape'?

Answer In primary hypersecretion of aldosterone, the ECF is not expanded without limit. Anti-aldosterone actions are activated so the body 'escapes' from uncontrolled sodium and water retention.

Concept The hormone aldosterone reduces urinary sodium excretion by stimulating active tubular sodium reabsorption.

If an adrenal cortex tumour secretes excessive aldosterone, a large positive sodium balance will result. This will be followed by an overexpansion of the ECF. There is a great isotonic expansion of the ECF.

Potentially, since the secretion of aldosterone is no longer feedback controlled, retention of sodium can progress. However, the body is triggered to increase natriuresis to oppose the aldosterone effects.

One anti-aldosterone contributor is the cardiac hormone atrial natriuretic peptide (ANP). As the name suggests, ANP increases filtered sodium load by increasing the glomerular filtration rate via renal arteriolar vasodilatation. The sodium reabsorption is also reduced either by any direct tubular epithelial actions or by blocking aldosterone actions at the nephron.

The kidneys also release a local natirureitc factor, called urodilatin to promote natriuresis. As a result of this anti-aldosterone activity, the ECF expansion is restricted and the development of edema is seldom encountered.

Within the kidneys, changes in renal hemodynamics during the hypervolemia also contribute to losing more of the excess sodium into urine. The renal sympathetic nerve that acts to conserve sodium is also inhibited via the baro- and volume reflex loops.

'Aldosterone escape' is not prominent when the stimulus of aldosterone release is due to a secondary cause. For example, in left cardiac failure, the reduction in effective circulating volume will be detected by intra-renal baroreceptors at the afferent arteriole. The enzymatic hormone, renin will be secreted. Sodium and water will be retained. There is expansion of the ECF and also peripheral edema.

Why does aldosterone action remain less opposed in this secondary cause of hyperaldosteronisms? Both circulating natriuretic and urodilatin should still be operative.

Perhaps the reduction in renal sympathetic action is not as much in heart failure as it is in primary aldosteronism that produces a ECF volume expansion.

This is because since the cardiac output is weak, the baro- and volume receptors will be sensing a lower blood pressure rise even though the blood volume is expanded.

4 How Do ADH and Aldosterone Integrated Actions Account for ECF Body Fluid Changes?

Answer Changes in sodium concentration will produce a corresponding stimulation or inhibition of ADH secretion. The new equilibrium ECF volume change will eventually be normalized by aldosterone action that restores and maintains the sodium balance.

Concept Osmolarity control is functionally related to sodium concentration in the ECF since the main determinant of ECF osmolarity is sodium and its compAnions. ECF volume homeostasis is under aldosterone hormonal control and ECF volume is tied to the total body sodium or sodium balance.

When dietary sodium input increases, before compensation by urinary excretion, the total sodium balance will be positive. This will produce an isotonic expansion of the ECF. The student should note that this isotonic expansion is mediated by the sensitive mechanism that maintains normal ECF osmolarity.

If we take this dynamic mechanism step by step, net dietary sodium input initially will raise the ECF sodium concentration. The osmoreceptors in the hypothalamus detects the increased sodium concentration/osmoalrity. The posterior pituitary responds with secretion of ADH to act on the kidneys to reabsorb water in order to normalize the ECf osmolarity.

Thus, the hyperosmotic plasma has led to the final isotonic expansion of the ECF by ADH.

Since the ECF and blood volume is above normal, eventually, euvolemia has to be achieved. The next regulatory step is the co-ordinated response of the body to the hypervolemia that resulted from the positive sodium balance.

Both neural and hormonal mechanisms to promote increased sodium excretion in the urine are activated. These include an inhibition of renal sympathetic activity (reflex from volume sensors), release of natriuretic peptides and inhibition of renin secretion.

The hypervolemia does inhibit ADH secretion but on its solo action, euvolemia will not be possible as any initial increase in urine volume with ADH suppression will increase the ECF osmolarity that will then trigger ADH release again.

The effective hormonal action that is needed is a reduction in plasma aldosterone level. The inhibition of renin serves this purpose. Tubular reasorption of sodium is decreased and hypernatriuria occurs in the hormonal answer to the positive sodium balance.

Dynamically, there is no peak natriuresis with aldosterone inhibition as is the case in a water diuresis with ADH inhibition. As the excess sodium is gradually lost, an initial decreased sodium concentration reduce the ADH secretion. More urine water is then excreted.

Thus the student must not imagine wrongly that since more sodium is not reabsorbed, water is secreted by the tubule, osmotically drawn by the sodium in the tubular fluid. There is no secretion of water to produce urine volume (this is a nineteenth century hypothesis, Wow Wee!). More urine volume is always due to less water reabsorption by the nephrons in the absence of ADH.

Thus we can see the hormonal duet between Aldosterone and ADH as they sing and play harmoniously about Osmolarity and Volume regulation. Aldosterone and ADH are a Homeostatic couple that look after their two 'so dium' Kids! (kidneys involved); the girl Concentration and the boy Balance.

5 How Does the Different Normal Concentrations of ECF Sodium and Potassium Fit in with the Sensitivity of the Mechanisms that Regulate the Two Cations?

Answer ECF sodium is high, at around 145 mmol/L while ECF potassium is maintained at a low normal concentration of around 4 mmol/L. The physio-logical mechanism that controls potassium homeostasis is therefore more sensitive directly to changes in plasma potassium than the mechanisms that govern sodium balance in the body.

Concept Since fluctuations in ECF potassium concentrations above or below its low controlled value (~4 mmol/L) can cause alterations in functions of excitable cells, the sensors must monitor well the potassium changes. The potassium sensors are located at the membranes of the endocrine cells in the adrenal cortex that secrete aldosterone. The membrane detection of potassium levels is direct.

Hyperkalemia will stimulate aldosterone secretion. In turn, the adrenocortical steroid hormone, aldosterone, will increase urinary excretion of potassium by promoting its secretion by the renal tubules.

Excess aldosterone action also increases proton secretion by the nephron. This produces an alkalosis. There is an interaction between blood pH and the plasma potassium level. This is mediated via a membrane 'exchange' phenomenon that is part of intra-cellular buffering in pH regulation. In alkalosis, there is a tendency for more renal tubular secretion of potassium resulting in hyperkaliuria and thus hypokalemia. Thus aldosterone has a direct and an indirect effect on ECF potassium.

For sodium at a much higher concentration of ~145 mmol/L, a 5 mmol/L change is insignificant. Thus sodium homeostasis is less 'urgent' and this is reflected in the more prolonged activation of the renin-angiotensin system that regulates total body sodium or the sodium balance.

That said, the student should be reminded that for the control of osmolarity which is determined by sodium concentration, small changes in ECF concentration are detected rapidly by hypothalamic osmoreceptors with a corresponding change in ADH (vasopressin) secretion from the posterior pituitary.

The big picture appears to be that if small changes in a solute affect cell function, the feedback control for that solute will also be sensitive to generate the necessary physiologic compensation. The concept of a need for direct, immediate sensing for solutes that are maintained at low concentrations also applies to plasma ECF calcium, which has an even lower concentration to potassium (~2.5 mmol/L). Calcium sensors are located on the plasma membranes of the endocrine cells that secrete parathyroid hormone. Small changes in calcium affect the membrane responsiveness of excitable cells.

For glucose with a normal low range at 4–6 mmol/L, the pancreatic beta cells directly sense the post-prandial hyperglycemia and respond rapidly to secrete insulin.

6 How Do Adrenaline and Noradrenaline Infusions Separately Affect the Cardiovascular Function Differently?

Answer Experimental infusion of adrenaline increases the cardiac output but for noradrenaline, a marked increase in blood pressure due to vasoconstriction produces a reflex bradycardia that accounts for a decrease in cardiac output.

Concept Both the adrenal medullary catecholamines, adrenaline and noradrenaline bind to both alpha and beta adrenergic receptors in various degrees. In vivo, plasma noradrenaline seldom exceeds its threshold for its observed cardiovascular (CVS) and metabolic effects. Most of noradrenaline effects are from its release as neurotransmitter from post-ganglionic sympathetic neurons.

In both experimental normal animals and in humans, slow infusion of noradrenaline and adrenaline leads to quite different CVS actions and the graphs shown in some Physiology textbooks often perplexed the students. In particular, the student knows that cardiac sympathetic action will always increase cardiac output via noradrenergic neurotransmission to both the sino atrial node and the ventricle muscles. The infusion effects of noradrenaline however show instead a reduced cardiac output.

Both adrenaline and noradrenaline bind to cardiac beta1 receptors to effect their chronotropic and inotropic actions on heart rate and myocardial contractility respectively. Noradenaline also has a strong alpha receptor action to increase the total peripheral resistance (TPR). The induced hypertension with noradrenaline (both systolic and diastolic pressure increase) triggers the baroreflex to produce a bradycardia.

The direct cardio-acceleratory effect of nordrenaline that is observed with cardiac sympathetic action is then overrided.

With adrenaline infusion, the total peripheral resistance is not markedly changed or slightly lower. This is due to the vasodilatory action of adrenaline on beta2 receptors on blood vessels in the skeletal muscles and in the liver (logical physio-logic action to increase blood flow to skeletal muscles and to promote hepatic glucose delivery during physical activity). Although adrenaline will also have alpha vasoconstrictor actions in other tissues, the net effect on TPR belongs to the beta2 binding.

Therefore with adrenaline, the pulse pressure is widen as the systolic pressure is increase but the diastolic pressure (affected by TPR) is decreased. There is no marked increase in arterial blood pressure to activate a baroreflex sufficient to produce a compensatory bradycardia. The direct actions of adrenaline on the beta1 cardiac receptors are unopposed and both heart rate and stroke volume are higher. Adrenaline causes the heart to pump out a greater cardiac output.

7 How Does Uncontrolled Diabetes Mellitus Affect the Potassium Plasma Level?

Answer The osmotic diuresis will lead to hyperkaliuria. In addition, the dehydration will further promote secretory loss of potassium by aldosterone that is increased by hypovolemia.

Concept Hyperglycemia in diabetes mellitus can overshoot the renal plasma threshold for glucose. The excess unreabsorbed glucose will interfere with the iso-osmotic reabsorption of water at the proximal tubule. Since filtered potassium is passively reabsorbed at the proximal tubule down its concentration gradient generated by prior water reabsorption, less filtered potassium will be recycled back to the circulation from the tubular fluid.

The osmotic diuresis will also affect the tubular event downstream at the ascending loop of Henle and at the collecting ducts. More water retained in the nephron will dilute the tubular fluid sodium and the activity of the triple sodium coupled symporter, Na/K /Cl at the ascending limb of Henle will also be reduced.

At the principal cells of the collecting ducts, the secretion of potassium will be enhanced by the greater tubular fluid flow, since the luminal second step in the trans-epithelial secretion of potassium is passive.

The hypovolemia resulting from increased excretion of urine volume will activate sodium conservation mechanisms which include the renin -angiotensin-aldosterone pathway. The aldosterone hormone action that recovers sodium will also stimulate tubular potassium secretion. Overall, kaliuria is increased by the osmotic diuresis in the diabetic patient, with resulting hypokalemia.

There are other factors that might potentially have hyperkalemic effects. The lack of insulin action will decrease the cellular uptake of potassium in all cells. There is some osmotic effect by the hyperosmotic plasma in diabetes to draw out some intracellular potassium. The dehydration itself from the polyuria will concentrate the plasma potassium.

Changes in pH can produce potassium shifts across cell membranes. This is part of intracellular buffering where protons are exchange for potassium cations. In diabetes mellitus with metabolic ketoacidosis, the keto anion can accompany the proton into the cell. Electroneutrality is maintained and there is no need for a potassium efflux.

In normal persons, the catecholamine adrenaline increases cellular uptake of potassium by a beta adrenergic effect. This is viewed as a useful action during physical activity to counter the exercise-induced hyperkalemia that is from the increased potassium efflux during action potential events. Exercise itself has an insulin-independent effect on cellular uptake of glucose and this provides the basis for recommending regular exercise for diabetic glucose control.

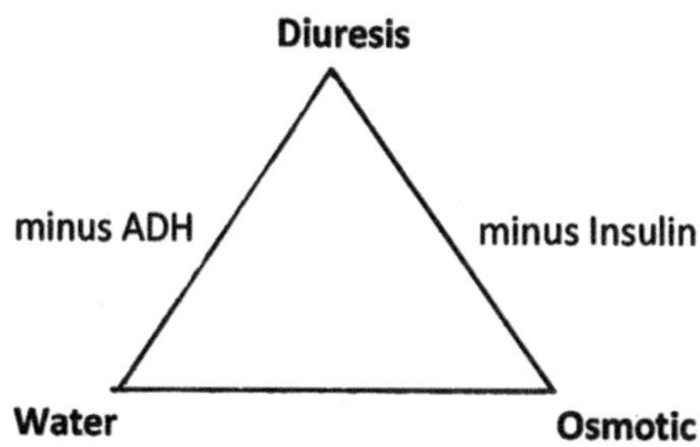

Increased urine volume excretion in two common endocrine disorders. In diabetes insipidus, lack of ADH action ('anti-dilute urine') causes inability to concentrate urine as the collecting ducts remain impermeable to water. In diabetes mellitus, increased filtered glucose load into the nephron interferes with water absorption by an osmotic effect, especially at the proximal tubules.

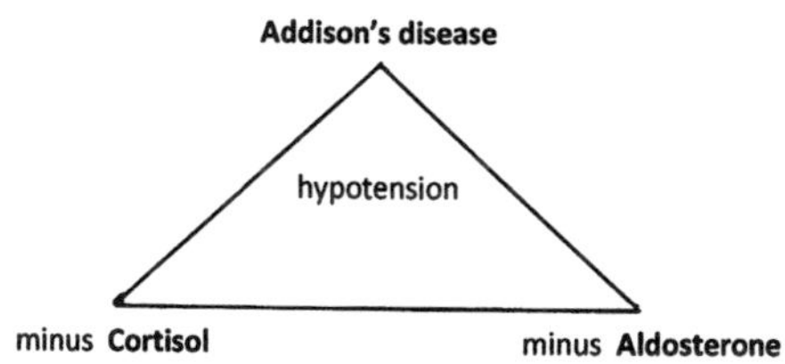

Primary adrenal insufficiency (Addison's disease) is a lack of all the three classes of adreno cortical steroid hormones. Little mineralocorticoid activity of aldosterone results in hypernatriuria and a negative sodium balance. There is hypovolemia. Added to this, the absence of an optimal cortisol action for normal vaso-responsiveness explains the decreased arterial blood pressure.

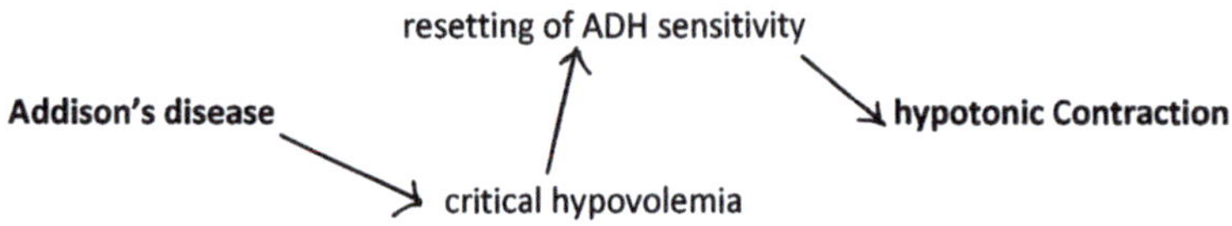

Hypotonic Contraction of the ECF is not an ECF disturbance seen in normal persons with intact ADH and Aldosterone mechanisms of response. One scenario when both the osmolarity and the volume of the ECF are reduced is in Addison's disease. This is due to the shift in priority when the severe hypovolemia threatens adequate cerebral perfusion. The volume sensitivity of the ADH response is thus reset and increased at the expense of a hypo-osmotic ECF.

Aldosterone and ADH are
a **Homeostatic couple**
that look after their
two 'so dium' Kids! (kidneys);
the girl Concentration and
the boy Balance.

Sodium physiology and the two 'A' hormones; Aldosterone and ADH.

When I am down,
Cortisol Increasing
When Finals Come,
High Adrenaline

My pancreatic
Insulin/Glucagon
Ratio Reduced
Sustain blood Energy

C-R-H Up
And I can stand the tension
A-C-T H
From my Pituitary

And I'll be Strong
Prof Cheng always behind me
Finals, I come
To Ace my Physiology!

Endocrine Song to the tune of 'You Raise Me Up"!

Glycemia	***Insulin***	***Hypo***
Hyper	***Secretion of***	***Leads to***

Make three possible physiologic cause and effect statements from these words (can omit one word per sentence). They are *'Hypo Secretion of Insulin Leads to HyperGlycemia' or 'Hyper Secretion of Insulin Leads to HypoGlycemia' and 'HyperGlycemia Leads to Secretion of Insulin' ('hypo' omitted).*

Chapter 9
Male and Female Physiology

The normal distinguishing, routine physiology in both sexes are basically hormonal. The testicular androgens and the ovarian oestrogen/progesterone have specific actions in males and females respectively.

The nervous system have no gender-related physiologic mechanisms. Neural communication are all action potentials. Some may argue for a female/male mind. During sexual excitement, there is a unique hard-wired, autonomic parasympathetic -induced penile vasodilation that achieves an erection for intercourse. Ejaculation to propel the sperms for potential fertilisation in the female tract is due to sympathetic action. The limbic/hypothalamus as well as thoughts, visual inputs, cutaneous sensory impulses bond the couple ecstatically during the sexual act.

During embryogenesis, the baby becomes female by default. In the absence of Mullerian inhibitory factor and testosterone from the fetal testis, the Wolffian ducts degenerate (men are wolfs!) and the Mullerian ducts differentiate into the female phenotype. Ex-utero, the baby grows, achieves adolescence and enters a reproductive cycle.

Interestingly, the female oestrogen is biosynthesized from local androgen precursors in the ovary. The other major source of androgens in females is from the adrenal gland. Testosterone acts directly and indirectly via a metabolite, dihydrotestosterone in target cells. There is also recent interest in the effects of testosterone in males acting through its cellular oestrogen metabolite.

The pregnant female during gestation secretes a variety of placental hormones that help to maintain the immuno-physiologic integrity of the placental trophoblast, the functional interface between mother and baby.

H.M. Cheng (ed.), *Physiology Question-Based Learning*,
DOI 10.1007/978-981-10-0877-1_9

1 Why Does the Anabolic Growth Hormone Act to Produce Catabolic Actions Like Glycogenolysis?

Answer The 'growth'-promoting effect of GH is related to protein sparing effects and in the adults, GH continues to have important metabolic effects in ensuring ongoing supply energy for sustaining health and tissue repair.

Concept The word 'growth' in GH should not be limited to merely linear bone growth in the child. Growth hormone secreted from the anterior pituitary is under a dual hypothalamic control, a stimulatory and an inhibitory releasing hormone. The latter GHIH, also called somatostatin is also found as a paracrine in gastric mucosa and in the pancreatic islet of Langerhans. Such a dual hypothalamic regulation will not make physio-economic sense if GH no longer has a role after the person stops growing in height.

Growth hormone is hyperglycemic and its actions serve to maintain blood glucose in situations of impending or present energy lack. For example, GH is increased during physical activity and in the fasting individual. GH acts to break down glycogen stores and GH also tends to reduce peripheral utilization of glucose. In conditions of energy demand, the maintenance of blood glucose ensures that the function of cerebral neurons that are very dependent primarily on glucose is not affected. The gastric hormone, ghrelin that is increased during the inter-digestive phase, when catabolism is needed, is named after ghrelin's effect on releasing GH. The hyperglycemia of GH actions is observed clinically in GH hypersecretion of acromegaly where the patient can excrete glucose in their urines, as the renal plasma threshold for glucose is exceeded.

The protein sparing effect of GH is also seen in its lipolytic actions. This metabolic action has been exploited in some wellness clinics that claim that injection of GH will help their clients to lose weight! GH increases fatty acid levels in the blood by its action on triglycerides in adipose tissues. The skeletal muscles can use the fatty acids as a metabolic fuel since cellular uptake of glucose is lowered by GH.

The diurnal profile of GH secretion also points to the growth and repair functions of GH. There is a peak rise in GH during deep sleep and this physiologic surge of GH during rest fits into the overall scheme of 'growth' hormone.

Some of the effects of GH on target tissues are indirect, via what was called initially 'somatomedins'. These include the insulin like growth factors, IGFs. The liver is a main source of the IGF when stimulated by GH. The cellular uptake of amino acids postprandially is increased by both insulin and GH-IGF actions. Feedback regulation of pituitary GH by IGF will be both on suppressing GH-releasing hormone (GHRH) and stimulating hypothalamic somatostatin.

The cellular protein synthesis machinery is stimulated by IGF-I. The IGF-II appears not to be under GH control and has a more predominant role in fetal development. Besides hepatic production, there is also local generation of IGF-I that has significant paracrine actions.

2 How Does Metabolism Affect Hormone Actions?

Answer Hormone actions are inactivated by metabolism after acting on target tissue to prevent prolonged excessive effects. Renal excretion also 'clears' hormones and their metabolites. Several hormones are transformed metabolically to their bioactive forms.

Concept The action of hormones are terminated soon after their physiologic actions. In general, peptide hormones (term used to encompass a 3 amino acid hypothalamic hormone to the large glycoproteins) and the catecholamines circulate unbound in blood. The total thyroid and steroid hormones in blood are predominantly transported bound to plasma proteins.

The bound hormones tend to persist longer in circulation as they are not filterable in the kidneys and not accessible to metabolic enzymes. Enzymes in the blood and target tissues deactivate rapidly the peptide hormones and catecholamines. The bioactive form of the bound hormones is the free species. The free active hormone is in equilibrium with the bound that also serves as a hormone reservoir. The free hormone can diffuse trans-capillary to bind to its target cells. For thyroid and steroid hormones, the dominant actions are effected via intra-cellular receptors, leading to gene activation and transcription with new protein synthesis.

The liver is the main metabolic factory for hormones. The cellular targets sometimes annul the hormone action by consuming the hormone-receptor complexes via endocytosis. Some hormone metabolites can be filtered or secreted, passively and actively in the kidneys. The shorter peptide hormones are freely filtered and the renal proximal tubular cells endocytose and catabolize the peptides to generate amino acids that are reabsorbed. The 'clearance' of hormones by the kidneys, pharmacologically defined, should be distinguished from the concept of 'renal clearance' used to describe renal handling of solutes and water (unit is volume of plasma/min not amount of solutes removed/min).

In a few important situations, metabolism activates and generates the potent hormone. We can think of the formation of angiotensin II from circulating angiotensin I by ACE (angiotensin-converting enzyme) in the lungs as blood courses along the ACE-endowed pulmonary endothelial cells. Thyroxine (T4) has its own direct actions but in many target cells, an intracellular de-iodinase produce T3 which then effects the actions of the amine hormone.

The male androgen, testosterone also has direct and indirect actions. In specific tissues, testotestone is first converted to dihydrotestoterone (DHT) by a reductase enzyme in the cells. The prostate gland is a example of a DHT target. Use of drugs that inhibit the reductase enzyme are prescribed in hypertrophy of the prostate.

Creation stories tell that the female was made from the male. Interestingly, in males, the female hormone estradiol is also a metabolite from testosterone in certain cells. This implies that the final observed physiologic actions in some male organs could be credited to estradiol! Students will recall that in the ovaries, local androgens provide the precursors for estrogen synthesis in the growing follicles.

3 What Are the Types of Endocrine Dysfunction Encountered?

Answer Endocrine failure or hypo-secretion can have primary glandular and secondary causes. Similar primary and secondary trophic hormone -related hyper-secretion also occur. Target cell hypo- or hyper-responsiveness in the presence of normal endocrine secretion is also encountered.

Concept When an endocrine gland secretes less than normal, this can be due to a variety of reasons. If the endocrine tissues are partially damaged, secretion will be reduced. An enzyme deficiency in the synthesis of the hormone e.g. steroid hormone is also a primary cause of hypo-secretion. For thyroid hormones, deficiency of a dietary factor, iodide leads to hypothyroidsm. Secondary hyposecretion is due to lack of the trophic hormone if the endocrine gland is e.g. under the hypothalamo-pituitary axis of control.

Primary and secondary causes of hyposecretion are distinguished by determining the level of the circulating trophic hormone. The stimulating trophic hormone is deficient in secondary causes but the trophic hormone is elevated in primary cases since the absence of negative feedback by the endocrine hormone raises the trophic factor.

Similarly, hyper-secretion can be primary, intra-glandular in origin or secondary, due to excessive trophic hormone action. If both the trophic and endocrine hormone concentrations in blood are elevated, a secondary hyper-secretion is present. A primary, hypersecreting endocrine gland will feedback negatively and lower the associated trophic hormone if there is a regulatory hormonal hierarchy involved.

Endocrine abnormalities can be associated with normal concentrations of the circulating hormones. Hypo-responsiveness occurs when there is deficiency or down-regulation of the hormone receptors e.g. in insulin resistance of diabetes mellitus. In obesity, leptin resistance appears to be present since leptin which increases metabolic rate is normal in the obese individuals.

If the receptor is abnormal and no longer binds the hormone, the target tissues are also non-responsive. We can think of nephrogenic diabetes insipidus where vasopressin receptors are altered. The kidneys cannot conserve water and a large volume, dilute (tasteless 'insipid') urine is excreted.

Interruption of post-receptor cell signaling pathways also accounts for cell unresponsiveness even when the hormone binds to its receptors.

Some active hormone species are generated by metabolism. For example, the testosterone bioactive metabolite, dihydrotestosterone is produced by an intracellular reductase in certain target cells of the androgens. A deficiency of the reductase results in loss of some of the secondary sex characteristics in the male (e.g. facial, body hair which might or might not be a secondary matter for the affected male!).

In cell hyper-responsiveness that is due to elevated thyroid hormone, the target tissues have increased effects due to the actions of a different hormone, i.e. catecholamines. This explains why the use of beta-blockers during a 'hyper-thyroid crisis' is effective in alleviating the physical symptoms like tachycardia.

The vascular hyperresponsiveness to vasoconstrictor hormones when there hyper-secretion of cortisol in Cushing's syndrome is another illustration.

In a different type of hormonal scenario, excessive glucocorticoid secretion causes an increased cellular effect in the renal tubules that is activated by mineralocorticoid receptor binding. There is sodium and water retention in Cushing's syndrome.

4 What Early Hormonal Actions Determine Male Differentiation in the Developing Fetus?

Answer The secretion of Mullerian-inhibiting factor and testosterone from the primordial fetal testes regresses the Mullerian ducts and activates development of the male internal genitalia respectively. The male external genitalia also requires the hormonal action of the metabolite, dihydrotestosterone.

Concept An essential generalization in sex differentiation ('genderalization'!) is that genes directly determine whether the fetus will develop testes or ovaries. The progression of sex differentiation depends in particular on hormones secreted by the testes.

Why (Y) are you male? The male and female gonads are derived from the same embryological tissue, the urogenital ridge. During the initial 6 weeks, the primordial gonads are asexual, undifferentiated. In the male, the Y chromosome has a sex-determining region (SRY) which is activated in the 7th week. The fetal testes are formed from embryonic cells in the urogenital ridge through a cascade of gene transcriptions triggered by the SRY gene product.

Before the fetal testes function, the primodial reproductive tract has a structure with a double genital duct (double d, the Wolffian ducts and Mullerian ducts), that can potentially develop into either the male or female phenotype. The fetal testes secrete two hormones that effect the sex differentiation of the internal and external male genitalia. One, the Sertoli cells in the testes secrete a developmental inhibitor, Mullerian inhibitory factor (MIF) that regresses the Mullerian ducts. MIF is a protein hormone.

Testosterone is concurrently secreted by the Leydig cells in the testes and the steroid hormone acts on the Wolffian ducts to produce differentiation into epididymis, vas deferens, ejaculatory duct and the seminal vesicles (remember, men are wolfs!). The external male genitalia develops under the action of the testosterone metabolite, dihydrotestosterone, produced in the target cells. The penis and scrotal tissues are formed and the testes descend into the scrotum . Deficiency of androgens in infants can cause failure of testes to descend (cryptorchidism) and infertility as spermatogenesis requires a temperature 2 °C lower than the body's core temperature.

The sex differentiation into female occurs in the XX fetus when the testes do not form. In the absence of testicular MIF and testosterone, the Mullerian ducts develop into fallopian tube, uterus, inner vagina, while the Wolffian ducts degenerate and outer vagina, external female genitalia form. The ovaries that are present in the 11th week of gestation thus do not contribute any hormones towards sex differentiation. The fetus will develop into a female by default unless the testes 'testify' hormonally!

The essential role of testosterone in male differentiation is seen in genetic male (XY) who have abnormal androgen receptors that result in androgen insensitivity syndrome. The testes are present and secrete MIF and testosterone. The Mullerian ducts degenerate but the Wolffian ducts are unresponsive to the testosterone. The 'wolf' does not grow! The formation of the male external genitalia that requires testostetone action is also not achieved and female genitalia and a vagina develop. There is cryptorchidism, usually discovered only at diagnosis. The abnormal sex differentiation in the phenotypic female (XY male; syndrome also called testicular feminization) is usually uncovered during consultation when 'she' visits the doctor due to failure of menstruation during 'her' pubertal age.

5 What Principal Placental Hormones Help to Ensure the Health of the Fetus and Mother?

Answer The placenta secretes human chorionic gonadotropin (hCG), placental growth hormones, progesterone, estrogen and corticotrophin-releasing hormone (CRH) during gestation. The fetal adrenal glands participate in the synthesis of the major placental estrogen, estriol.

Concept The hCG is produced by the placental trophoblast as early as 6 days post-implantation of the fertilized ovum. The hCG has a structural similarity of luteinizing hormone (LH) and acts primarily by binding to target LH receptors. hCG rescues the corpus luteum from programmed cell degeneration. This ensures the continual production of luteal progesterone and estrogen until the placental steroidogenesis sufficiently takes over, to sustain the feto-placental allograft. hCG peaks at 2–3 months and maternal hCG blood level is a useful monitor for the placental trophoblast health.

The placenta produces human placental lactogen (hPL), now called human chorionic sommatomammotropin (hCS). A variant of growth hormone is synthesized by the placenta and is the dominant circulating GH in maternal circulation from the 5th month of gestation. The maternal pituitary GH becomes undetectable probably due to feedback inhibition from the placental GH. This feto-placental GH increases lipolysis and decreases glucose utilization in the mother to divert glucose for the fetus. Interestingly, the placental GH is not measurable in the fetal circulation, thus fetal growth is determined by other growth factors.

The placenta is unable to synthesize cholesterol from acetate. Therefore placental progesterone is made from cholesterol in circulating low density lipoproteins. Progesterone is needed for maintaining uterine quiescence for a successful pregnancy to term. Progesterone achieves this partly by inhibiting uterine prostaglandin production. The unique survival of the feto-placental graft is also due to immunomodulatory action of progesterone.

The placental synthesis of estrogen involves an interesting interplay between the fetal and maternal adrenal glands . The placenta lacks the 17 alpha-hydroxylase enzyme and so cannot convert progesterone to estrogen or to synthesize androgens. This latter deficiency in placental androgens is designed to protect the female fetus from potential masculinization. In addition there is active placental aromatize activity that transforms maternal and fetal androgens to estrogens. The placenta extracts 16-hydroxylated DHEA (dehydroepiandrosterone from the fetal adrenals) from the fetal circulation and produces the major estrogen, estriol. Principal actions of estrogen include increasing uterine growth, development of the mammary glands and inhibition of lactation.

Parturition is timed by a 'placental clock'. The placenta also secretes a corticotrophin-releasing hormone (CRH) that has similar structure/function to the hypothalamic CRH that regulates adrenal cortisol release. Placental CRH progressively increases during pregnancy and peaks during the onset of labour. Fetal hypothalamus also secretes CRH. Since cortisol is required for maturation of the fetal lung and synthesis of pulmonary surfactant, the fetus appears to self-timed physiologically its entry into the world, when the 'placental lung' in her watery environment is replaced by air-filled respiratory alveoli.

6 How Are Several Positive Feedback Mechanisms Involved in the Reproductive Life of a Mother?

Answer The mechanism of ovulation and follicular growth are mediated by a positive action of oestrogen. Parturition and the suckling reflex are also positive feedback neuro-endocrine loops.

Concept The hypothalamo-pituitary-ovarian axis regulates the endocrine secretion of oestrogen and progesterone. Classical feedback inhibitory control of the ovarian steroid hormones exert effects on the anterior pituitary and hypothalamus by respectively decreasing gonadotrophin (FSH, LH) and GnRH secretions.

In the ovarian cycle, at a temporal point just before ovulation and rupture release of the ovum, a reverse unusual positive stimulation of pituitary luteinizing hormone secretion is caused by the rising oestrogen levels as the 'chosen' follicle grows and enlarges. The stimulus trigger of ovulation is the surge in LH blood levels. The follicle secretes oestrogen under the action of FSH. The precursor for oestrogen synthesis is local androgen production stimulated by LH (male hormone needed to produce the female kind!). The oestrogen has an autocrine positive action in increasing the number of FSH receptors in the growing follicle.

When parturition is triggered, initial uterine smooth muscle contraction activates an autonomic reflex. The mechano-receptors are stimulated and afferent impulses relayed to the hypothalamus . The hypothalamic neurohormone, oxytocin is released from the posterior pituitary. The increased blood oxytocin levels cause stronger uterine contractions during labour, activating also nociceptors. The greater contractile activity continues to prolong the mechanoreceptor -oxytocin-uterine wall positive feedback response to aid in the birth of the baby. The elevated levels of oestrogen during late pregnancy can also upregulate the oxytocin receptors on the uterine smooth muscle.

The happy mother nurses her baby and a positive neuroendocrine reflex ensures the contentment of the baby. When the baby sucks on the nipple, tactile, sensory receptors are stimulated and this leads to generation of afferent signals. The suckling reflex turns on the secretion of oxytocin and the hormone acts on the myoepithelial cells in the mammary glands to eject milk for the baby's feed. The supply of milk on suckling demand encourages the baby to continue the mechanical oral milking at the mother's nipple. Breast-feeding also has the beneficial effects in helping the stretched-out uterus to return towards its pre-pregnant size via oxytocin contractile actions.

Concurrently, the supply of milk for the baby's subsequent feed (like sms, sensory message 'top-up'!) is ensured by a parallel neuro-endocrine reflex that releases the hormone prolactin from the anterior pituitary. This neuroendocrine reflex, also triggered from the nipple receptors, could disinhibit the major inhibitory hypothalamic control factor (dopamine) on pituitary prolactin secretion. The nursing baby indeed positively massages both the posterior and anterior pituitary for its own nutritional needs.

If both breasts are suckled simultaneously, as occurs during the feeding of twins, more prolactin is released. Both the prolactin and oxytocin neuroendocrine reflexes have afferent neural pathways that travel via the spinal cord and brainstem enroute to the hypothalamus.

7 Is the Penile Erection Response an Autonomic or Somatic Reflex?

Answer The spinal autonomic reflexes that produce penile erection are both facilitated and inhibited by descending cortical-spinal descending fibers.

Concept The erectile tissues of the penis are basically venous sinusoids that are surrounded by strong fibrous tissues. Erection is a haemodynamic event that is triggered by parasympathetic nerve action (erection during sexual excitement is not a 'fight or flight reaction!). The parasympathetic fibers travel in the pelvic nerve with the cavernous nerve as the final effector that causes vasodilation of the internal pudendal artery and its branches.

At rest or in absence of sexual excitement, cerebrally or cutaneously, the sympathetic vasoconstrictor input is dominant. During erection, a nonadrenergic, noncholinergic parasympathetic autonomic activity takes over. The neurotransmitter is nitric oxide which stimulates the enzyme guanylyl cyclase that generates cyclic GMP. This second messenger leads to relaxation of the vascular smooth muscles that sees an acute inflow of blood that produces the penile engorgement. Specific phosphodiesterase inhibitor drugs allow the cGMP to persist and is used to treat erectile dysfunction to help sustain and proLONG the erection. The vascular expansion also passively compresses the veins that empty the penile circulation and this enhances the penile rigidity.

The autonomic reflex vasodilation of the penile arterioles is unique in vascular physiology as the resistance of most other arterioles that control regional blood flow is altered by the degree of sympathetic adrenergic constrictor fibers, i.e. vasodilation is effected by a reduction in the sympathetic activity. The secretion of lubricating mucus by the bulbourethral glands during sexual arousal is also parasympathetic-mediated and helps prepare for impending coitus.

Sexual thoughts or genital stimulation produce penile erection. Psychogenic reasons can also inhibit adequate penile sexual response required for penetration. The afferent fibers from the genital 'erogenous'areas are relayed in the pudendal nerve to the sacral spinal cord. The efferent parasympathetic fibers run in the pelvic nerve. Erection is also called tumescence and detumescence is likely contributed by a reversal to sympathetic activity.

While erection is a parasympathetic event, ejaculation is powered by sympathetic action. The smooth muscles of the epidydymus, vas deferens, ejaculatory ducts, prostate and seminal vesicles all contract with sympathetic stimulation. This achieves emission to empty the semen (sperm and glandular secretions) into the urethra. Expulsion of the semen is accomplished by strong contractions of the urethral smooth muscles as well as skeletal muscles at the base of the penis.

The penile erection is thus an integrated autonomic/somatic response. Erection has both voluntary and involuntary neural components. The eLongation and hardening of the penis is an automatic and yet conscious physiological event. During sleep, there is also the phenomenon of nocturnal emission ('wet dream').

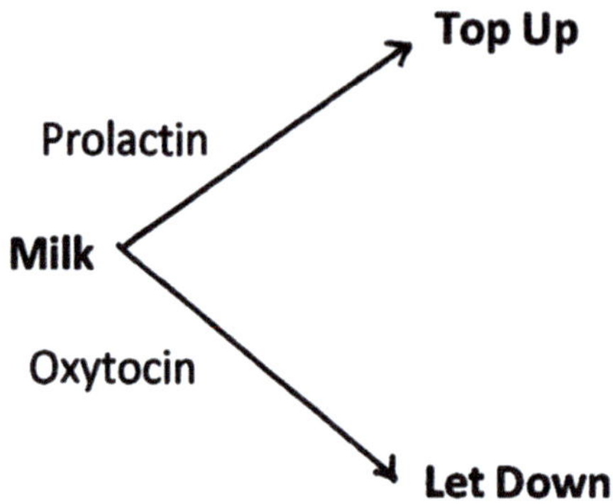

Both anterior and posterior pituitary are involved to satisfy the baby's nutrition. The secretion of prolactin and oxytocin are both increased by neuroendocrine reflexes arising from the stimulation of sensory receptors in the nipple when the baby suckles on the mother's breast.

Gender Humour

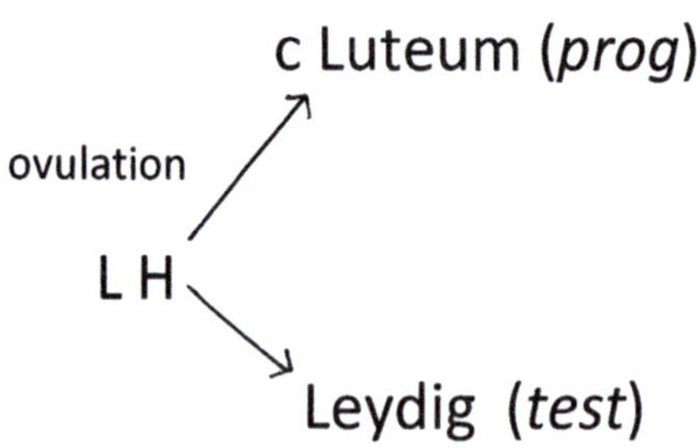

Luteinizing hormone (LH) from the anterior pituitary is released by hypothalamic gonadotrophin releasing hormone. LH surge is the stimulus for ovulation and that 'egg-traordinary' event begins the luteal phase of the ovarian cycle. The corpus luteum is stimulated by LH to secrete Progesterone to sustain uterine development in preparation for a potential pregnancy. If fertilization is absent, there is a programmed cell death of the corpus luteum, culminating in menstruation. In males, LH stimulates the Leydig cells to secrete the androgen, testosterone.

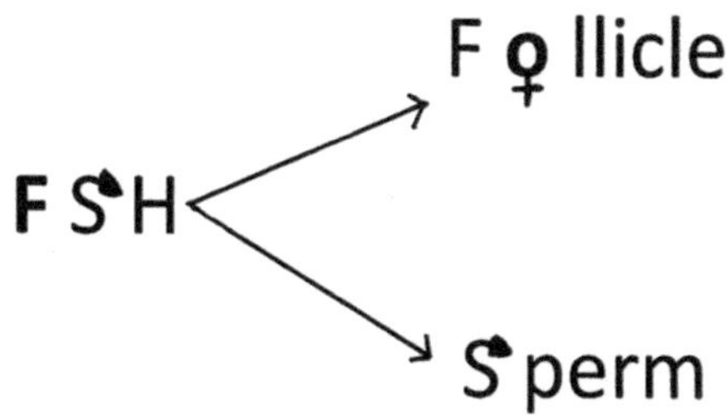

The anterior pituitary, follicle stimulating hormone (FSH) is released by hypothalamic gonadotrophin releasing hormone (GnRH). FSH is essential in both female and male physiology. FSH stimulates the growth and maturation of the ovarian follicles which secrete oestrogen. Spermatogenesis requires FSH which acts together with testosterone.

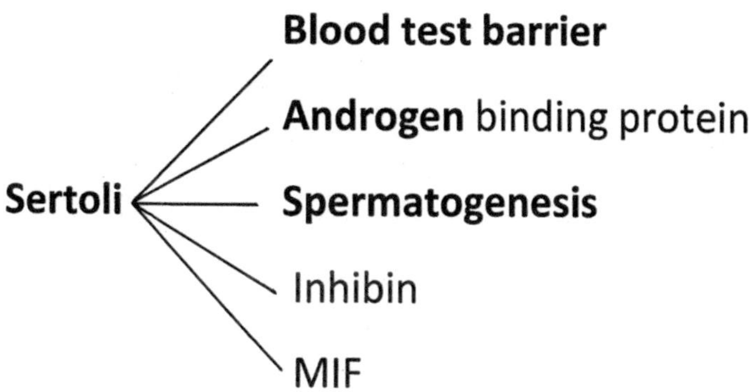

The Sertoli Cells (SC) of the seminiferous epithelium in the testis are 'certainli' important in their diverse actions to ensure normal spermatogenesis. The SC histologically establish the blood-testis barrier. The androgen binding protein concentrates testosterone to promote sperm maturation. Inhibin is a negative feedback regulator of FSH. Mullerian Inhibitory factor (MIF) determines male sexual differentiation during embryogenesis.

Part III
Gastrointestinal System

"I am sending Onesimus back to you, sending my very heart (*splanchna*)" Letter to Philemon

Introduction: You Are What You Eat

The common term 'Digestive tract' from the mouth to the anus is spatially outside the body proper. The Bronchial tree in the respiratory sysktem is also exposed to the air in the atmosphere. Thus the mucosal immune system has an essential physiologic role in the gastrointestinal tract and in the lungs to protect the human host from invasion by microorganism. In recent years, there is increasing awareness of the gut microflora/microbiota and their microbio-physiological contribution to health, including in immunity and energy balance.

The oral and gastrointestinal system (GI) function like a elongated mechanical processor. Food is broken down and progressively digested along the tract up to the small intestine.

Absorption of the final enzymatic products of dietary carbohydrates, lipids and proteins take s place in the small intestines. Undigested food residues are dessicated in the large intestines with further absorption of water.

The normal GI motility is a one-way traffic and this is ensured by the co-ordinated activity of the enteric neural system (ENS) of interconnecting neurons that is part of the histological architecture of the tract. The ENS receives input from both the parasympathetic and sympathetic nerves. Together the intrinsic ENS and the extrinsic autonomic nervous system (ANS) act to achieve the sequential secretory and motility patterns of the different GI segments.

The central nervous system (CNS) is also involved in the behaviour of the GI tract. The ENS in the gut is sometimes called the little brain. The phenomenon of nervous diarrhea illustrates the CNS effects on the GI physiology. The brain-gut axis is a bidirectional relationship. During a meal, when the body is enriched with energy, hormones from the GI tract acts on the hypothalamus to mediate sensation of satiety.

The GI tract is not a silent organ during the inter-digestive or fasting period. Periodic peristalsis activity is observed, the migrating, myoelectric motility complex (MMC). The MMC serves a 'housekeeping ' function to clear food debri and prepare the gut for the entry of the next meal. The hormone motilin contributes to MMC and hunger sensation is also an action of the gastric hormone ghrelin, which is increased between meals.

Specific membrane transporters are localised on the luminal and the basolateral sides of the enterocytes. This polarized intestinal epithelial cell ensures a unidirectional absorption of glucose, amino acids, small peptides from the lumen to the capillary. Most of the products of lipid digestion take a different absorptive pathway, involving micelles and chylomicrons, through the enterocytes via the lymphatics and eventually into the systemic circulation.

Move, Digest, Absorb, Excrete. These four primary activities of the GI tract proceed automatically once masticated food is swallowed several times a day. Muscles, Enzymes, Hormones, Neurons all come into play at designated time and place along the GI tract. The GI system is designed for optimal food digestion and absorption of energy nutrients. The Cardio-respiratory system supplies the oxygen and removes metabolic CO_2 while the GI functions provide the energy substrates for cell metabolism.

Chapter 10
Motility and Peristalsis

"*...bowels of mercies...*" Epistle to the Colossians

In Greek, the bowels (*splanchnon*), is also thought of as the site of affections and feelings. The splanchnic circulation that supplies the gastrointestinal (GI) tract derives its name from this Greek word. We talk about bowel movements and e-motions was believed to be from this GI source.

Motility is an automatic event of the GI tract smooth muscle (except for the upper one-third of esophageal skeletal muscle). The motor activity is progressively one-way and this uni-directional movement is ensured by the enteric nervous system (ENS). The ENS is hard wired into the bulk of the multilayered longitudinal, circular smooth muscle/connective tissue of the GI.

The networking of neurons in the ENS is modulated by the autonomic parasympathetic and sympathetic inputs. Parasympathetic is the dominant autonomic discharge during feeding and digestion. The central nervous system (CNS) is therefore interlinked with the GI physiology via the autonomic nervous system (ANS) which powers or depresses the ENS effects on secreto-motor activities. Just the sight of a KFC Zinger poster will zing some central neural impulses that then travel to the GI structures.

The GI motility resembles a mechanical processor. There is mixing and fragmenting action of the different patterns of GI motility. Along the GI elongated tract, there are smooth muscle sphincters that serve to control the direction and rate of the flow of chewed food and chyme. The upper and lower esophageal sphincters, the pylorus, the ileo-cecal and anal sphincters are the major tolls or regulatory stops along the GI highway.

The chymic flow rate along the small and large intestines are also neurally metered. Too fast a transit time results in hypermotility diarrhea as insufficient time is available for water reabsorption. Conversely, depressed intestinal motility causes constipation.

H.M. Cheng (ed.), *Physiology Question-Based Learning*,
DOI 10.1007/978-981-10-0877-1_10

1 Esophageal Function

Elaborate on these facts concerning esophagus during physiologic swallowing.

a. The vagus nerves form motor endplates in the upper third of the esophageal muscle wall.
b. Pressure in the thoracic esophagus is close to intrathoracic pressure.
c. Lower esophageal sphincter tone is controlled by vagal excitatory fibers and vagal inhibitory fibers.
d. A secondary peristalsis happens if primary peristalsis does not completely clear the esophagus of food.
e. The esophageal phase of swallowing is controlled primarily by the swallowing center.

The esophagus contains a gradient of muscle, from all skeletal at the top to all smooth at the bottom. The esophageal musculature is mainly innervated by branches of the vagus nerve. Motor fibers of the vagus nerve form motor end plates on the striated muscle cells. Visceral motor nerves (preganglionic parasympathetic) make connections mainly with neurons of the myenteric plexus. The resting pressure in the lower esophageal sphincter (LES) is about 20 mmHg. Vagal cholinergic activity contributes to this basal tone. This prevents reflux of gastric content and the development of esophagitis (heartburn). Reflux is a potential problem because the intraesophageal pressure is almost always less than intraabdominal pressure.

Primary peristalsis is controlled by the swallowing center located in the medulla and lower pons. Esophageal peristalsis leads to a reflex relaxation of the LES, involving vagal inhibitory fibers to the circular muscle of the LES. After primary peristalsis, a secondary peristalsis can be triggered due to distension of the esophagus by uncleared food.

The afferent limb of the swallowing reflex is initiated when touch receptors, especially in the pharynx are stimulated. In the pharyngeal phase, the upper esophageal sphincter relaxes and a peristaltic wave is generated with contraction of the pharyngeal constrictor muscles. In the esophageal phase, after the bolus of food passes the UES, the sphincter reflexly constricts. Primary peristalsis then begins just below the UES.

2 Enteric Nervous System (ENS)

Respond and elaborate on the following descriptions on the intrinsic ENS.

a. Mucosal sensory neurons release predominantly CGRP into interneurons in the enteric plexuses.
b. Quantitatively, NO is the most important mediator of GI smooth muscle relaxation.
c. Most mucosal sensory cells respond indirectly to sensory stimuli. Comment.

d. Localized mechanical or chemical stimulation of intestinal mucosal does not cause bilateral contraction on each side of the point of stimulation. Comment.
e. Vasoactive intestinal peptide has excitatory effects on smooth muscle blood vessels and gland cells. Is this correct?

The submucosal ganglia have numerous sensory neurons. These neurons are the afferent limbs of secretomotor reflexes. Impulses in the sensory neurons are mostly, indirectly evoked by mechanical or chemical stimuli. Enterochromaffin cells (EC cells) in the mucosa release serotonin (5-HT) in response to chemical or mechanical activation. Sensory neurons are then stimulated by serotonin. The sensory neurons project to enteric ganglia and dorsal root ganglia. The major neurotransmitter that binds to interneurons in the enteric plexuses is calcitonin-gene-related peptide (CGRP). In contrast, the myenteric stretch receptors respond directly to stretch. Submucosal ganglia also contain vasodilator neurons that release acetylcholine and/or VIP onto submucosal blood vessels. The neurotransmitter VIP has an excitatory secretomotor effect in epithelial and gland cells.

Most neurons in the myenteric ganglia are motor neurons. These include both excitatory and inhibitory neurons. Local stimuli characteristically produces contraction above (oral) and relaxation below (anal) the site of stimulation of the GI tract. Excitatory neurons release acetylcholine and substance P and nitric oxide (NO). There is also a positive interaction between VIP and NO.

Most myenteric interneurons release acetylcholine onto nicotinic receptors on motor neurons or on other interneurons.

3 Gastrointestinal Motility

Comment on why the following statements are **incorrect.**

a. There is approximately one million neurons In the GI tract.
b. The frequency of slow wave in the GI tract is relatively constant.
c. The amplitude of slow wave does not change.
d. Smooth muscle cannot contract without action potential.
e. Action potentials produce smooth muscle twitches.

About the same number of neurons (100×10^6), found in the spinal cord constitute the semiautonomous enteric nervous system (ENS). The ENS comprises motor neurons, interneurons and sensory receptors. The extrinsic parasympathetic and sympathetic nerves project primarily onto neurons of the myenteric and submucosal plexus of the ENS. In this manner, the extrinsic innervation modulates the motor and secretory activities of the GI tract via the ENS.

The frequency of the slow wave (basic electrical rhythm) varies from about 3 per minute in the stomach to 12 per minutes in the duodenum. The amplitude of the BER can be influenced by intrinsic and extrinsic nerves and by hormones and paracrine mediators. If the peak of the slow waves exceeds the firing threshold, a train of action potential (1 to 10/s) is triggered. The greater the number of action potentials, the more intense is the smooth muscle contraction. Slow waves that are unaccompanied by action potentials elicit weak contractions of the GI smooth muscle cells. Smooth muscle cells contract much slowly (about one-tenth that of skeletal muscle). Thus, the individual contractions in a train of action potentials summate temporally. This produces a gradual, increasing smooth muscle tension.

The smooth muscle cells, particularly the circular layer of the GI tract are electrically-coupled by many gap junctions.

4 Gastrointestinal (GI) Motility

Comment on the following events of GI motility.

a. The gastroileal reflex is a long-range reflex.
b. Motilin is needed for generation of migrating myoelectric complex.
c. Antipropulsive movements occur in the proximal colon.
d. The colonic circular muscle seldom fires action potentials.
e. Mass movement occurs in the colon.

The stomach and the ileum interact in a reflex called the gastroileal reflex. Increased secretomotor activity of the stomach increase the motility of the terminal ileum and speed up the movement of luminal contents through the ileaocecal sphincter. Distension of the ileum activates neural reflexes and release of peptide YY that reduce gastric emptying.

The MMC in the stomach is generated by vagal impulses that release motilin. In the duodenum and small intestine, the MMC is independent of extrinsic nerves and requires motilin. The MMC also inhibits the migration of colonic bacteria into the terminal ileum. Poor MMC contractions may lead to bacterial growth in the ileum. Bacterial substances stimulate the secreto-motor activity in the ileum and thus results in diarrhoea. Most contractions in the cecum and proximal colon are segmental (haustration). Large portions of the colon are involved in haustration. Reverse peristalsis and segmental propulsion towards the cecum both take place in the proximal colon. Chyme is retained in the proximal colon to provide for adequate absorption of electrolytes and water.

Acetylcholine increases the duration of the slow waves. The lengthening of the slow waves elicits contractions of the circular muscle. Besides slow waves, the colon also exhibits low amplitude, high frequency myenteric potential oscillations.

About one to three times daily, a wave of a prolonged contraction propel the luminal contents of the colon over a significant length in an orthograde direction. In the gastrocolic reflex, after the entry of food onto the stomach, the frequency of mass movement is increased.

5 Gastric Contractions/Emptying

Respond and elaborate on the following aspects of gastric motility.

a. Can contractions occur without action potential?
b. Acetylcholine and gastrin both stimulate gastric contractility.
c. Gastric emptying is slowed by duodenal pH below 3.5.
d. Gastric emptying is decreased by hypertonic chyme.
e. Does the hormone gastrin affect the lower esophageal sphincter?

The frequency of gastric slow wave is about 3/min, generated by a pacemaker site near the mid-body of the stomach. The gastric low wave resemble the triphasic cardiac action potential. However, the gastric slow wave is ten times larger and does not overshoot. Gastric action potentials frequently occur during the plateau phase of the slow wave, mainly in the terminal antrum and pylorus. Contractions produced by the action potentials are much stronger than those that occurs without action potentials. Both acetylcholine and gastrin increase the amplitude and plateau phase of the gastric slow wave. Noradrenaline has the reverse effect. The longer the muscle cell remains above the threshold for contraction and the greater the depolarization, the bigger is the contractile force.

The gastroduodenal junction permits carefully controlled gastric emptying so that the duodenum can further process the chime at an effective rate. The junction also prevents regurgitation of duodenal contents as the gastrin mucosa can be damaged by bile. The duodenal and jejenal mucosa have receptors that respond to acidity, osmolarity, fat digestion products and peptides/amino acids. Chyme is often hypertonic and further duodenal digestion also increases the osmoactive particles. Both neural and hormonal factors are involved in regulating gastric emptying. For example, acid in the duodenal activate neural reflexes and hormone secretin secretion. Antral contractions are inhibited while pyloric sphincter contraction is increased. Thus acid is not released into the acid-sensitive duodenal mucosa more rapidly than it can be neutralized by pancreatic and duodenal secretions. Large doses of gastrin increase the tone of the LES. However, this effect is not seen at the circulating gastrin levels after a meal.

6 Gastric Motility

Comment on the inaccuracy of the following statements on gastric motor functions.

a. When food enters the stomach, intense contractions and mixing proceed throughout.
b. Gastric mixing occurs immediately with food in the stomach.
c. Peristaltic wave in the antrum relaxes the pyloric sphincter.
d. In the fasted person, there is no antral contraction.
e. The migrating myoelectric complex (MMC) affects the stomach through to the colon.

Distension of the esophagus and esophageal peristalsis relaxes the LES and this is followed by receptive relaxation of the fundus and body of the stomach. The muscularis externa of the fundus and body is relatively thin. These parts of the stomach serve a reservoir function with little increase in intragastric pressure. Vagovagal reflex mediates the receptive relaxation. Gastric contents may remain unmixed here for 1 h after entering. Fats tend to form an oily layer as the contents settle down based on their densities. As a result, fats have a slower rate of gastric emptying.

The major mixing function occurs in the antrum. Antral peristalsis reflexly constrict the pylorus and the chyme is emptied in small squirts, one for each peristalsis wave. Food particles greater than 2 mm do not pass the narrow pyloric opening. The retropulsion at the terminal antrum helps to mix and break down the gastric contents.

In the water-digestive period, a cyclic contractile activity (MMC) is observed in the antrum and sweeps from the stomach to the terminal ileum. The antrum is quiescent for 75–90 min and then a 5–10 min of electrical and motor activity takes place. This serve a housekeeping role to clear even large portions of food material that remain from the previous meal. The strong periodic antral contractions occur with a relaxed pylorus.

7 Explain the Role of the Following Factors in Gastric Emptying

a. Fats
b. Cholecystokinin
c. Gastric inhibitory peptide
d. Amino acids
e. Gastrin

Presence of fat-digestion products in the duodenum and jejunum decreases the rate of gastric emptying. Partly, the release of CCK from the duodenum slows the gastric emptying. Fatty acids also stimulate the release of the hormone, GIP, which also causes constriction of the pyloric sphincter. The hormone CCK also stimulates duodenal neurons that trigger vagovagal reflexes that decrease gastric emptying. Thus the rate at which fat in the chyme is emptied into the duodenum is regulated so that it does not exceed the rate at which the fat is emulsified. This is by the bile acids and lecithin of the bile.

Peptides and amino acids release gastrin from G cells, present in the antrum of the stomach and the duodenum. Gastrin increases gastric antral contractions and also pyloric sphincter constriction. Overall gastric emptying is diminished. Peptides and amino acids also cause the release of GIP and CCK from the duodenum. In duodenal ulcers, the mechanisms that release gastric emptying-reducing hormones from the duodenum may be ineffective.

Hypertonic solutions in the duodenum elicits an unidentified hormone that lowers the rate of gastric emptying. In the absence of the stomach, excessive ingestion of food can result in a 'dumping syndrome' which is partly due to the hyperosmolarity of duodenal contents.

Duodenal chemoreceptors thus regulate gastric emptying through hormonal mediators as well as via intramural intrinsic plexuses.

8 Small Intestinal Motility

Elaborate on the following smooth muscle activity of the small intestines.

a. The duodenum makes up 5 % of the small intestine.
b. The most frequent type of movement is segmentation.
c. The frequency of slow waves is highest in the duodenum.
d. Smooth muscle excitability is inhibited by sympathetic nerves.
e. Maximal rate of segmental contractions is the same as the frequency of slow waves.

The duodenum has no mesentery. The jejunum occupies 40 % of the length of the small intestine. Most of the digestion and absorption takes place in the duodenum and jejunum.

The frequency of slow waves decreases from about 13/min in the duodenum to about 9/min in the terminal ileum. Action potential bursts on the slow waves are localized to short segments of the intestine. This produces the rhythmic segmentation which effectively mixes the chyme with digestive secretions and facilitates the contact of chyme with the mucosal surface. Segmentation is closely spaced contractions of the circular muscle layer. The sites of circular contractions alternate so that individual small neighbouring segments of the gut contract and then relax reciprocally. The frequency of the action potential spikes depends on the excitability of the smooth muscle. The parasympathetic nerves enhance the excitability and this effect is effected via the intramural plexuses.

Peristaltic waves occur only over a short length of the intestine (short-range peristalsis). The slow propulsive movement of chyme allows adequate time for digestion and absorption. The small bowel is about 5 m in length and chyme normally takes 2–4 h to transverse it. Castor oil, a laxative contains hydroxyl fatty acids that stimulate small intestinal motility. The intestinal transit time is reduced. Salts and water are delivered to the colon at a rate that surpasses the colonic absorptive capacity. Diarrhoea is produced.

9 Jejenum and Large Intestines

a. Why is the jejunum relatively sterile?
b. What happens to indigestible carbohydrate?
c. What is the effect of dietary fiber in the large intestine?

d. What is the tonicity of chyme along the GI tract?
e. How does the colon secrete potassium?

Water moves to and fro across the intestinal epithelium until the osmotic pressure of the intestinal contents equal that in plasma. Depending on the nature of the ingested meal, the duodenal contents can be hypertonic or hypotonic. By the time the chymic material enter the jejunum, its osmolarity is close that of plasma. This isotonicity is maintained throughout the small intestine. When the osmoactive nutrients are absorbed, water follows down the osmotic gradient generated by the transepithelial transfer of the digested food products. Saline cathartics like magnesium sulfate acts as an osmotic diarrhoeal agent, exerting a laxative effect.

The chyme in the jejunum is relatively free of bacteria. The ileum has some but the colon normally contains large numbers of bacteria. The lingering presence of gastric acid and the more rapid chymic movement through the jejunum may inhibit the bacterial growth.

One function of intestinal flora is the formation of indigestible carbohydrates. Short-chain fatty acids (SCFA) are produced, along with several gases which contribute to flatus. SCFA seem to be trophic to the colonic mucosa and stimulates sodium absorption. Cellulose, hemicellulose and lignin are components of dietary fibers. Dietary fiber adds bulk to the GI contents. It exerts a hygroscopic effect so that the stools are bulkier and more easily excreted. A shorter mouth to anus transit time is said to reduce the risk of colon and rectal carcinomia. A reduction in intestinal contact time by bacterial toxins and injurious metabolites could be part of the reason.

The mechanism for colonic K^+ secretion is similar to that in the principal cells of the late distal tubule and the collecting ducts. Aldosterone also has an action here on K^+ secretion and Na^+ absorption. Flow rate enhanced K^+ secretion and loss is seen in diarrhea.

10 Anus, Rectal Canal

Respond and elaborate on the following characteristics of the terminal end of the GI tract.

a. Filling of the rectum occurs by a mass movement.
b. Distention of the rectum results in reflex relaxation of the internal anal sphincter.
c. Reflex relaxation requires the presence of enteric neurons.
d. How long do the reflex responses continue for?
e. During defecation, what is the role of the sympathetic neurons?

Just before defecation, a mass movement in the sigmoid colon fills the rectum. The stretch of the rectum reflexly relaxes the internal anal sphincter and also causes reflex constriction of the external anal sphincter. Involuntary defecation is prevented because of the functional motor nerves. The reflex sphincter reactions are transient. If defecation is not done immediately, the sphincters recover their normal tone and the urge to defecate is temporarily diminished.

In congenital megacolon (Hirschprung's disease), frequently, the enteric neurons are lacking in the internal anal sphincter and a short stretch of the neighboring colon. Since reflex relaxation does not occur, obstruction and dilation of the colon above the obstruction results.

The major efferent nerves during defecation are cholinergic parasympathetic fibers is the pelvic nerves. Sympathetic activity does not have a significant role. The integrating neurons for the defecation reflex are located in the sacral spinal cord and they are subjected to higher center neural influences. Propulsive contractions of the descending and sigmoid colon take place during the defecation reflex.

Voluntary actions relaxes the external anal sphincter. Voluntary effort may raise the intraabdominal pressure to as high as 200 cm H_2O.

HOMEOSTASIS, the word coined by Walter Cannon, describes the feedback/compensation mechanisms that maintain a constancy of the ECF, the 'milieu interieur' of all cells, as conceptualized by Claude Bernard. Homeostatic motor activity and motility. The neural and endocrine mechanisms that undergird the GI pattern of smooth muscle activity involve several feedback loops e.g., esophageal secondary peristalsis, gastric emptying and migrating myoelectric complex, activated at different stages of GI functions.

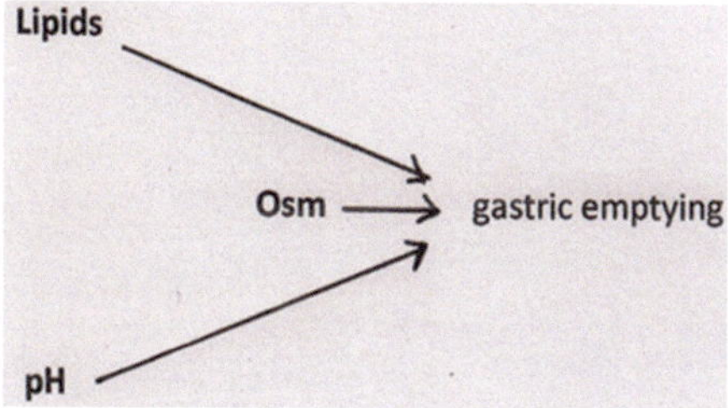

Gastric release of chyme is a controlled event effected by neuro-endocrine mechanisms, involving hormones like CCK, secretin and local enteric reflexes. The characteristics of the chyme self-regulate the gastric emptying. A lipid-laden meal empties slower to allow time for duodenal processing of fats by bile and lipolysis by pancreatic enzymes. Low acidity also slows gastric emptying to ensure sufficient neutralization by pancreatic bicarbonate. Hyperosmotic chyme inhibits emptying to prevent occurrence of a reverse flux of water into the lumen. This is the scenario in 'dumping syndrome' when gastric storage is absent and overeating inappropriately loads food directly into the intestines.

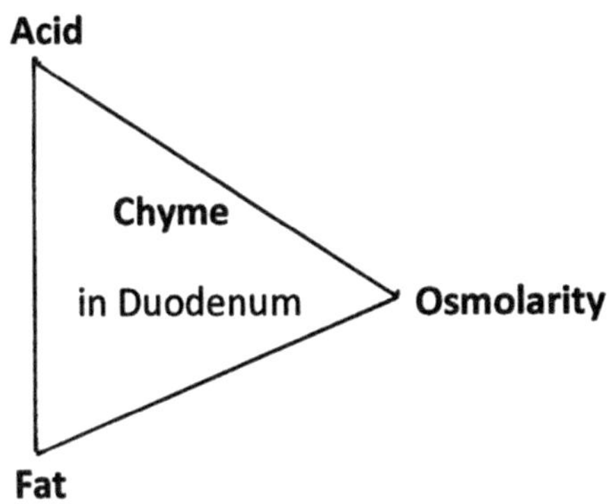

The three major characteristics of gastric chyme that self-regulate gastric emptying.

Chapter 11
Secretion of Digestive Juices

During a meal, a large volume of digestive juices include the salivary secretions, the gastric juice, the pancreatic and bile secretions. Daily, the total volume of exocrine gland secretions into the gastrointestinal lumen (7 plus liters) is easily three times larger than the water intake. Most of the water of the digestive juices are reabsorbed by the small and large intestines.

This implies that a large increase in blood flow occurs concurrently to the various exocrine glands during stimulated secretions. The aqueous portion of the secretions are derived from the plasma of the blood capillaries that perfuse the glands. Parasympathetic nerve activity increases the salivary and splanchnic blood flow.

Except for salivation which is basically neurally regulated, the gastric, pancreatic and bile secretions are under neuro-endocrine controls The various secretions are timed and sequentially released in concert with the presence of masticated food in the mouth and arrival of the food bolus or chyme at different segments of the GI tract.

Besides digestive enzymes, the digestive juices also has protective roles. The salivary bicarbonate and the pancreatic bicarbonate neutralize acidity and prevents acidophilic bacteria growth orally and duodenal erosion respectively. Gastric low acidity is also bactericidal. An adjacent gastric mucosal secreted layer of bicarbonate and mucus also contribute to the gastric barrier to prevent acidic tissue injury and autodigestion.

Hepatic bile, secreted and stored in the gall bladder is released during digestion to enable the digestion and absorption of fats in the aqueous milieu of the GI lumen. Amphipathic or amphiphilic bile salts are responsi**BILE** for successful fat processing during a meal.

1 Saliva

Elaborate and respond to the following concerning saliva.

a. Amylase is active between pH 4 and 11.

H.M. Cheng (ed.), *Physiology Question-Based Learning*,
DOI 10.1007/978-981-10-0877-1_11

b. The largest salivary gland secretes little mucins.
c. Maximal blood flow to salivary glands is greater than an equal mass of active skeletal muscle. Comment.
d. What is the tonicity of saliva?
e. The potassium concentration is always much greater than in plasma.

Salivary amylase has the same enzymatic specificity as pancreatic amylase. In the stomach, amylase continues to operate until antral mixing lowers the pH to below 4.0. If chewing is substantial, more than half of starch can be broken down to oligosaccharides.

The parotid glands, the largest salivary glands, are entirely serous. The submandibular and sublingual are mixed mucous and serous glands. Serous acinar cells secrete amylase found in zymogen granules at the apical side of the cell. Maximal rate of saliva production in human is about one ml/min/g. This is equivalent to the gland's weight/min. Blood flow is proportional to the high rate of salivary secretion. Maximal perfusion to salivary glands is about tenfold than in the same weight of contracting skeletal muscle. The parasympathetic nerve innervations release acetylcholine and VIP, both of which produce the vasodilation.

Saliva is always hypotonic to plasma, about 70 % that of plasma. The primary secretion is isotonic. The salivary ducts remove more Na^+ and Cl^- than they add K^+ and HCO_3^-. The faster the salivary flow rate, the nearer to isotonicity it will be. Resting saliva is slightly acidic. With increasing salivary flow, the saliva becomes basic, up to pH 8.0. This is partly due to the greater bicarbonate concentration with higher flow rate.

The salivary K^+ concentration with higher flow rate is very low. The ducts only modify the composition of the primary salivary secretion. The volume of saliva is not affected.

2 Salivary Secretion

a. What is the primary physiological control of salivary glands?
b. How does cyclic AMP affect secretion?
c. What is the role of calcium in serous acinar cell secretion?
d. What is the effect of noradrenaline?
e. How is chloride absorbed by the salivary ducts?

Most of the gastrointestinal secretions are controlled hormonally. In contrast, the major regulator of salivary secretion is the parasympathetic fibers in the facial and glossopharyngeal nerves (cranial nerves VII and IX respectively). Parasympathetic stimulation increases the synthesis and secretion of amylase, mucins. The neural activity also increases ductal transport processes, elevates blood flow as well as enhancing glandular metabolism and growth. Sympathetic activity also increases, salivary flows but the parasympathetic effects are greater and more prolonged.

Noradrenaline acts on alpha and beta receptors to increase cytosolic calcium and cAMP respectively.

Effectors that elevate cAMP result in a primary secretion that is richer in amylase. This includes the action of VIP. Agonists that increase intracellular calcium lead to a greater volume of acinar cell secretion but with a lower concentration of amylase. Substance P and acetylcholine produce this response in the serous acinar cell.

Chloride enters by a HCO_3^-/Cl^- anion antiporter at the apical membrane. Basolateral K^+ channels that are activated by increased cytosolic Ca^{++} maintain the electronegativity of the cytosol. This helps to drive chloride absorption across the basolateral membrane via a chloride channel. The chloride conductance may be enhanced by increased cytosolic cAMP or Ca^{++}.

The impermeability of the ductal epithelium to water leads to the hypotonicity of the saliva as the ducts modify the salivary composition by a net reabsorption of electrolytes.

3 Gastric Secretion

a. Why does the pH of the gastric contents decrease when chyme moves to the small intestine?
b. What effect has atropine on the vagal stimulation of gastric HCl secretion?
c. Why does cimetidine have a greater HCl inhibitory effect than expected?
d. Why can bet H^+ secretory rate be lower in gastric ulcer?
e. Why is the parietal mass increased in Zollinger-Ellison Syndrome?

Gastric HCl is inhibited when HCl is no longer required to activate pepsinogen to pepsin. This occurs when the gastric chyme has moved to the small intestine. Decreased pH gastric content is the main inhibitory regulator of HCl secretion. The reason is due to the buffering capacity of food. When food is emptied from the stomach, the gastric pH decreases and thus inhibits gastrin release. In addition, somatostatin also inhibits gastrin secretion. Somatostatin is released by acid in the lumen and it probably acts in a paracrine way via gastric juice to suppress gastrin secretion.

Atropine will block the cholinergic stimulation of parietal cells. However, atropine will not inhibit the vagal action on gastrin secretion since the neurotransmitter is not acetylcholine but GRP (gastrin-releasing polypeptide).

The three agonists of parietal cells (acetylcholine, histamine and gastrin) also potentiate each other's effect. The combined response is greater than the sum of the individual responses. Thus, the H_2 receptor blocker, cimetidine produces a bigger HCl inhibitory response than expected. Cimetidine blocks the direct action of histamine as well as the histamine-potentiated effects.

In gastric ulcer, the gastric mucosal barrier is damaged. This then permits the H^+ and pepsin to autodigest the mucosa. Net H^+ secretion can be lower due to the leak of secreted H^+ into the defective mucosa. Gastrin secretion is elevated due to the reduced inhibition by gastric H^+.

H^+ secretion is greatly elevated in gastrinoma (commonly in the pancreas). Gastrin also has a trophic effect on the parietal cells. Duodenal ulcers develop and acid-inactivation of pancreatic lipases result in steatorrhoea.

4 Gastric Juice Secretion

a. What are the three physiological agonists of HCl secretion by parietal cells?
b. What happens to parietal cell when the basolateral K^+ channels are activated?
c. What happens to the luminal membrane chloride conductance?
d. What stimuli release intrinsic factor?
e. What is the strongest agonist of HCl secretion?

Histamine, gastric and acetylcholine are the three physiological activators of parietal cell HCl secretion. Each of these is an example of paracrine, endocrine and neurocrine control of GI secretions. Acetylcholine and gastrin elevate the cytosolic calcium concentration while histamine increases the intracellular cAMP.

There are two types of potassium channels in the basolateral membranes of the parietal cell. One type of K^+ channel is Ca^{++}-activated and the other type is activated by cAMP. Greater efflux of K^+ through these channels hyperpolarizes the parietal cell. This increases the driving force that promotes chloride anion movement out of the cell through the apical electrogenic Cl^- channels. The chloride conductance during secretion is also markedly increased by the raised cytosolic Ca^{++} and cAMP concentrations. Furthermore, cAMP and Ca^{++} also stimulate the insertion of more Cl^- channels into the luminal membrane. The number of H^+/K^+ ATPase in the canalicular membrane is also increased. This is effected by the secretory canaliculi. During HCl secretion, the parietal cells undergo a morphological change. Extensive membrane fusion occurs to increase the number of HCl-extension sites.

The same stimuli that produce HCl secretion also release intrinsic factor from the parietal cell. Intrinsic factor is the only gastric component that is essential for life.

Histamine has greater potency on parietal cells than acetycholine and gastrin. Much of the response to gastrin is a consequence of gastrin-activated release of histamine from ECL cells. Histamine H_2 receptor blockers can reduce the parietal HCl response that is due to gastrin.

5 Gastric Acid Secretion

Elaborate on the following regarding gastric parietal cell functions.

a. The gastric glands contain enterochromaffin-like cells.
b. The major anion of gastric juice is chloride.

c. The H^+ concentration gradient across the parietal cell membrane is about one million-fold.
d. Chloride exchanges with bicarbonate across the basolateral membrane.
e. Secretion of H^+ requires an ATPase.

The gastric mucosa can be classified into three distinct regions. In a gastric gland, the mucous neck cells can differentiate into columnar epithelial cells. The parietal and peptic cell (secrete HCl and pepsinogens respectively) are located deeper in the gland. The oxyntic glands also have ECL cells that secrete histamine and D cells that release somatostatin.

Potassium is always higher in gastric juice than in plasma. Thus severe vomiting can precipitate hypokalemia. Chloride is the major anion at all rates of gastric secretion. At higher rates of secretion, gastric juice is like an isotonic solution of HCl. The pH in parietal cell cytosol is 7.0 while the gastric lumen is about 1.0. This corresponds to a million-fold gradient against which the parietal cell must secrete H^+.

The apical membrane of the parietal cell contains a H^+/K^+ ATPase exchanger. Benzimidazole drugs like omeprazole reacts irreversibly with the sulfhydryl groups of H^+/K^+ ATPase at the low pH.

When H^+ is actively pumped out of the parietal cell, there is an excess of gradient across the basolateral membrane. Chloride influx into the cell across the basolateral membrane is driven and energized by the downhill efflux of HCO_3^- via a HCO_3^-/Cl^- countertransporter. The lumen of the stomach is electronegative by 30–80 mV relative to the serosa. Thus chloride enters the gastric lumen against both chemical and electrical potential differences.

6 Gastric Acid Regulation

Respond and elaborate on the following aspects of gastric acid physiology.

a. The most important endogenous antagonist is somatostatin.
b. In the cephalic phase, is the direct vagal input to the parietal cells the only stimulus for acid secretion?
c. How much acid can be secreted during a cephalic phase?
d. The acidity of gastric contents regulates itself? Explain.
e. What effect does intact proteins have on HCl secretion?

Somatostatin directly inhibits HCl secretion by parietal cell. Somatostatin also decreases gastrin secretion from the G cells. The cephalic phasc is evoked by the sight, smell and taste of food before the food enters the stomach. The vagal fibers stimulate enteric neurons that are mainly cholinergic. These cholinergic neurons directly produce the cephalic phase gastric HCl secretion. Indirectly, the acetylcholine also increases acid secretion by releasing gastrin from antral G cells and histamine from gastric ECL cells.

In the absence of food, the pH of the antral contents drops rapidly during the cephalic phase. Acid secretion during this period may be up to 40 % of the total secreted during a meal. However inhibitory mechanisms activated by low pH in the antrum limits the amount of acid secreted. When the gastric mucosa is bathed with a solution of pH 2.0 or less, HCl secretion by any agonist mechanism is effectively blocked. The inhibition is brought about by direct action on parietal cells, local inhibitory neural reflexes and reduction of gastrin from G cells.

The gastric phase of gastric secretion is initiated by the presence of food in the stomach. The dominant stimuli are distention of the stomach and presence of protein digestion products (peptides and amino acids). Amino acids, particularly phenylalanine and tryptophan, and peptides cause antral G cells to release gastrin. Distention triggers local and central, vagovagal reflexes via mechanoceptors stimulation. Gastric distention enhances the chemical activation of parietal HCl secretion.

7 Gastric Mucosal Barriers

Elaborate on the following protective parameters of the gastric mucosal barrier.

a. Gastric mucins are about 80 % carbohydrate.
b. The stomach has a very thin layer of viscous, alkaline coat.
c. Gastric secretions can transverse the mucus layer.
d. The pH of the gastric epithelial cell surface is maintained at about 7.0.
e. Non-steroidal inflammatory agents reduce the efficacy of the gastric mucosal barrier.

Mucins are secreted by Mucous neck cells. About 80 % of a gastric mucin is carbohydrate. Mucin consists of four monomers, each 500×10^3 daltons in weight. The protective mucus layer requires continuous production of tetrameric mucins to replace those cleaved by pepsins. The glycosylated part of the mucin is resistant to proteolysis. Mucus production is stimulated by some of the same stimuli that increase acid and pepsinogen secretion. Acetycholine is especially important. Mechanical deformation of the gastric mucosa also elicits neural reflexes that promote mucus secretion.

Bicarbonate secreted is entrapped by the viscous mucus HCO_3^- makes the mucus layer alkaline. Bicarbonate secretion is enhanced by parasympathetic acetylcholine release. The mucus layer is about 0.2 mm thick, and it effectively separates the HCO_3^- at the surface epithelial cells from the acidic contents of the gastric lumen.

There are 5–7 μm diameter channels through the mucus layer. Through these, the gastric secretions can move. The mucus allows the pH of the epithelial surface to be maintained at around 7.0, while the gastric luminal pH is about 2.0. Gastric mucosal barrier requires both the mucus and bicarbonate secretions.

Drugs like aspirin inhibit both the secretion of mucus and HCO_3^- secretion. This may result in the pathogenesis of stress ulcers.

8 What Is the Role of the Following Factors in the Intestinal Phase of Gastric Secretion?

a. Duodenal distension.
b. Peptides and amino acids.
c. Bulbogastrone.
d. Hypertonic chyme.
e. Low pH of chyme.

The entry of chyme in the duodenum activates neural and endocrine responses that initially stimulate and later inhibit acid secretion in the stomach. This depends on the activity of the gastric chyme. When the buffering capacity of the gastric chyme is exceeded and the pH of the emptied chyme falls below 3.0, inhibition predominates.

Duodenal distention increases gastric acid secretion by vagovagal reflexes that then stimulate the antral G and parietal cells. There are also G cells in the duodenum and proximal jejunum. Peptides/amino acids stimulate gastrin release from these cells. In addition, absorbed amino acids and peptides may circulate to the antral G cells to enhance further gastrin secretion.

Intestinal hormones that influence gastric secretions are called enterogastrones. Acid I the duodenal bulb stimulates bulbogastrone which reduces acid secretion by the parietal; cells. Acid in the duodenum also suppresses parietal cell HCl release via enteric and vagovagal reflexes. Additional inhibition is through the release of the hormone secretin.

Fat digestion products in the duodenum and proximal jejunum inhibits HCl secretion via the hormones cholecystokinin and GIP. An unestablished enterogastrone partly mediates the gastric acid reduction produced by hypertonicity in the duodenum.

9 Pancreatic Secretion

Respond to the following regarding the exocrine function of the pancreas

a. What is the principal source of HCO_3^- secreted into the lumen?
b. During which phase does most of the secretion occur?
c. How does CCK potentiate the action of the hormone secretion?
d. The major direct agonist of acinar cell is acetylcholine.
e. What is the nature of the juice secreted during the gastric phase?

Bicarbonate in the perfusing blood of the pancreas is the main source of the secreted bicarbonate. The mechanism is similar to the renal proximal tubular reabsorption of HCO_3^-. It involved secretion of H^+ by the basolateral membrane Na^+/H^+ exchanger and H^+/K^+ ATPase. The CO_2 formed from the reaction between secreted

H^+ and HCO_3^- enters the ductal epithelial cells and is hydrated, catalyzed by carbonic anhydrase. Dissociation of the carbonic acid produces HCO_3^- and H^+. The HCO_3^- enters the lumen via the HCO_3^-/Cl^- exchanger.

At rest, the aqueous component is made primarily by the intercalated and intralobular ducts. When secretin hormone is increased, the additional aqueous component is produced mainly by the extralobular ducts. Secretagogues that elevate cAMP potentiate the effects of those that raise intracellular Ca^{++}. Secretin elevates cytosolic cAMP. Apparently, the more important effect of CCK in stimulating acinar cell enzyme-rich secretion is indirect CCK stimulates the afferent arms of vagovagal reflexes that in turn trigger secretion by acinar cells and ductal cells. In response to a meal, about 65 % of the total pancreatic juice is produced when chyme comes into contact with the duodenum and upper jejunum. In a single day, the pancreas secretes about ten times its mass of pancreatic juice.

Distension of the stomach evokes vagovagal and enteric gastro pancreatic reflexes that stimulate acinar and duct cells. The juice produced is small in volume with a high enzyme concentration.

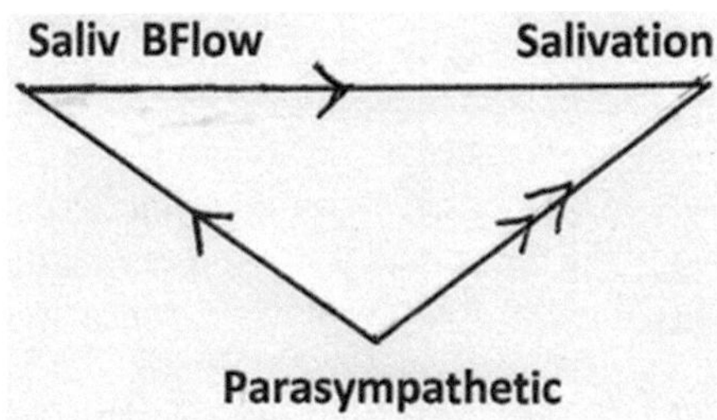

The regulation of salivation, during cepahlic, conditioned reflex or oral phase is mediated predominantly by parasympathetic nerve. The parasympathetics increase simultaneously, the salivary flow and the blood flow to the salivary glands. The aqueous volume in the saliva is derived or secreted from the plasma.

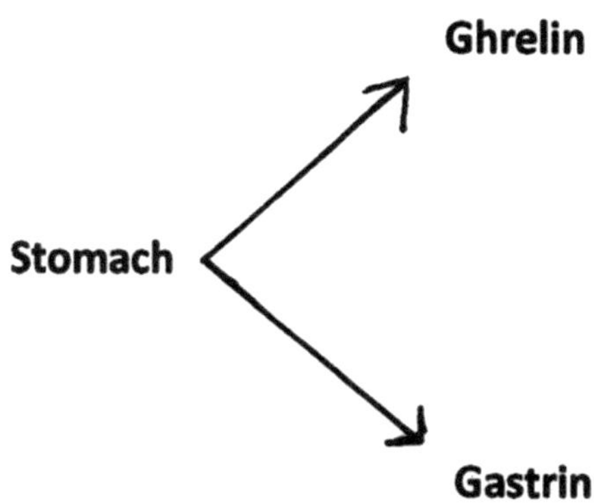

The stomach secretes two major hormones, gastrin during a meal (gastrin) while ghrelin is increased during the inter-digestive period and it stimulates appetite. The name ghrelin is coined from its action in stimulating growth hormone (GH) release. GH is hyperglycemic and ghrelin's effect on GH during fasting does have some physiologic sense. Gastrin increases acid secretion that promotes protein digestion by pepsin and eventually, the amino acid absorbed into the blood stimulates GH secretion, which increases cellular uptake of the amino acids.

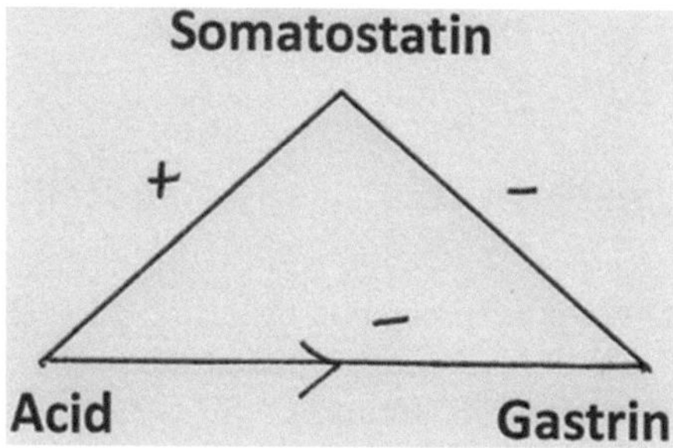

The gastric paracrine somatostatin is the major 'brake' on the secretion of acid. This action of somatostatin is indirect by suppressing gastrin release and this removes a major stimulus on the acid-secreting gastric parietal cells. The low intra-luminal gastric acidity in an empty stomach stimulates somatostatin paracrine secretion.

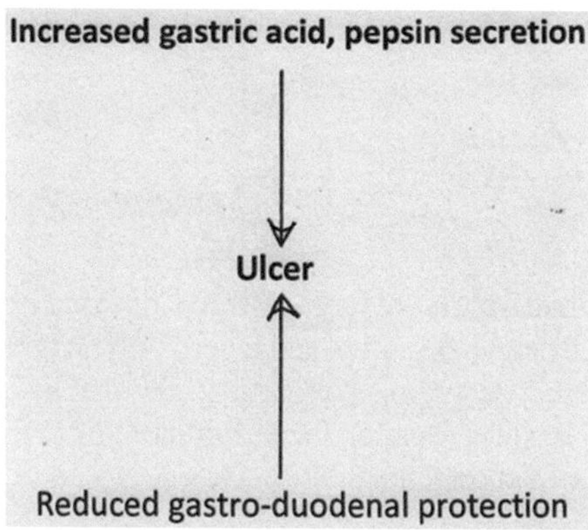

The incidence of GI ulceration is dependent on the balance between gastric proteolytic activities in acidic milieu and the tissue self-protective, structural and functional mechanisms that shelter gastric and duodenal cells from autodigestion and acidic tissue corrosion.

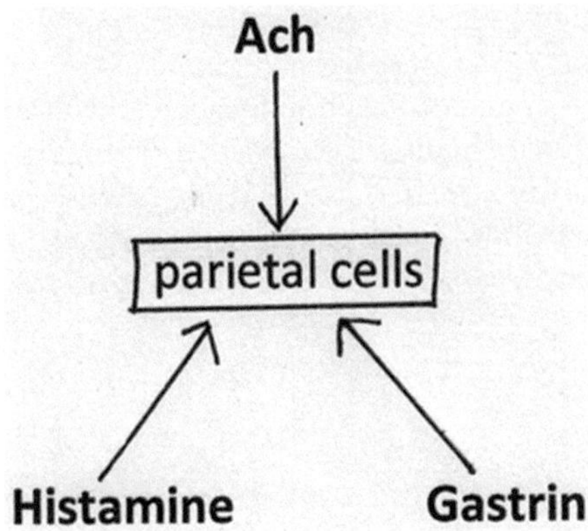

The three major stimuli of proton secretion by gastric parietal cells. Acetylcholine is a neurotransmitter at both ENS neurons and at the post-ganglionic parasympathetic nerve fibers that innervate the stomach.

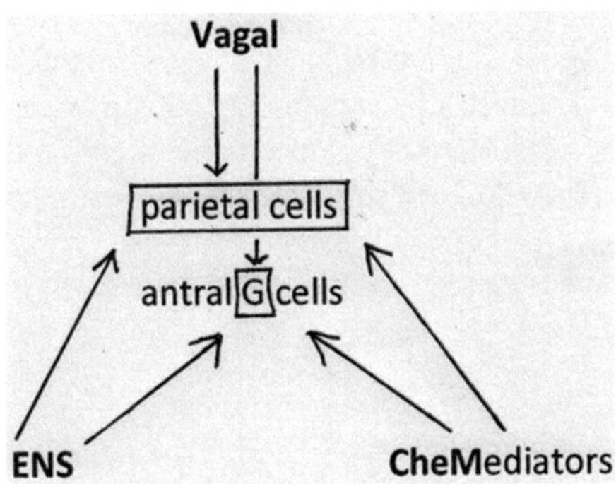

The antral G cells secrete gastrin. In turn, gastrin is released by specific vagal nerve activity, by stretch-activated ENS reflex and by proteolytic products in the stomach. The parietal cells that secrete hydrogen ions are innervated by excitatory enteric and vagal neural inputs. Besides gastrin, local mediator like gastric histamine also increases hydrochloric acid secretion.

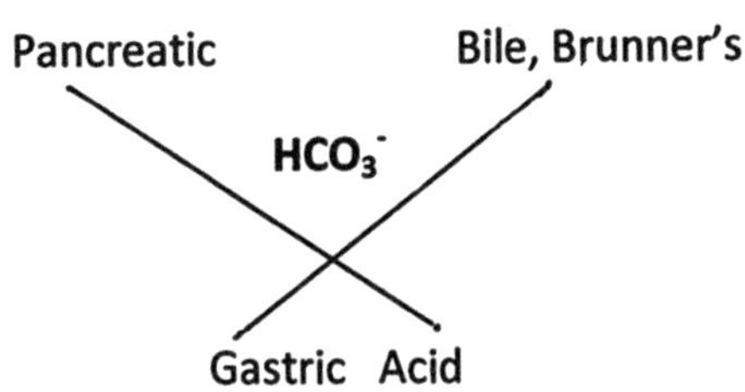

Neutralization of the acidic gastric chyme is needed as the pancreatic enzymes act at a higher pH in the duodenal lumen. The base, bicarbonate is secreted in the pancreatic juice, bile and from intestinal Brunner's glands.

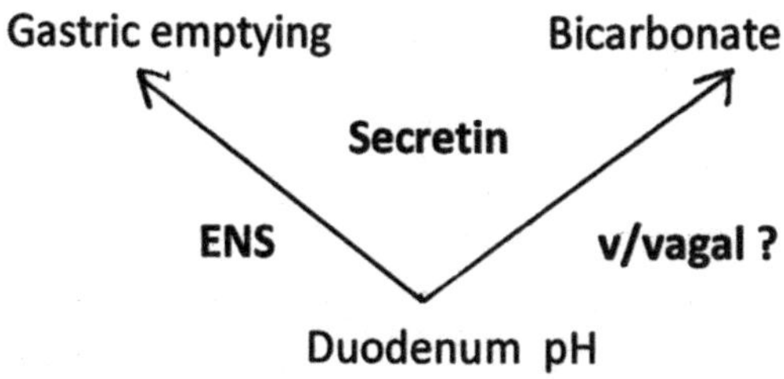

The duodenal hormone secretin (first named 'hormone') is part of two feedback loops, one with the pancreas and the other with the stomach. The trigger for both loops is duodenal pH. Secretin increases the bicarbonate content of pancreatic juice. Secretin also slows down the entry of acidic chyme from the stomach. Besides the mediator role of secretin released in response to pH, there are possibly also local enteric and long loop vagal reflexes, activated by pH that govern gastric emptying and the bicarbonate secretion.

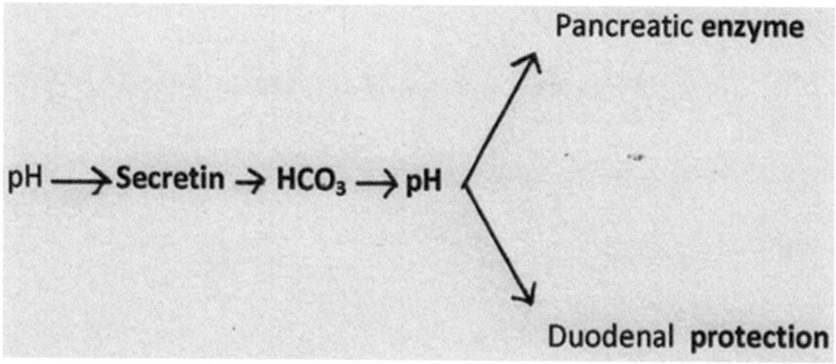

Duodenal pH and secretion of the hormone secretin forms a feedback loop. The release of a bicarbonate rich pancreatic juice by secretin when acidic chyme touches the duodenal wall accomplishes two purposes. The acid neutralization raises the luminal pH for optimal pancreatic enzyme action. Secondly, the duodenum, unlike the stomach, has no definite 'mucosal barrier' and needs to be protected from acidic injury.

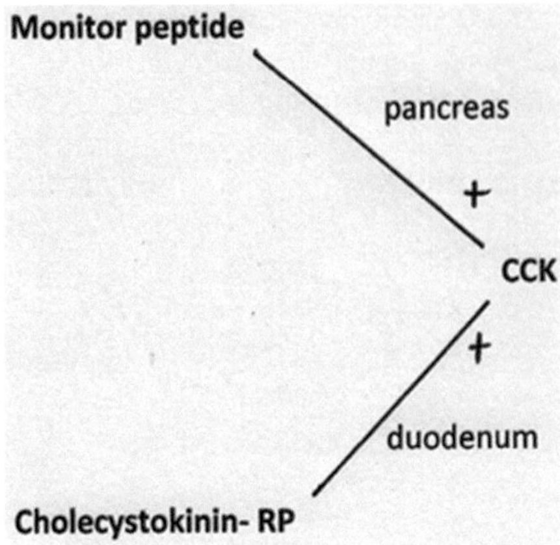

Recent studies on cholecystokinin action have revealed two regulatory peptides that regulate CCK release. Monitor peptide is produced by pancreatic acinar cells, while CCK releasing peptide is from the duodenal cells. The two peptides help to match pancreatic secretion of proteolytic enzymes to the requirement for these proteases in the intestinal lumen. When the proteins in the meal are completely digested and absorbed, both these control peptides will themselves be sacrificially degraded by the pancreatic enzymes.

Chapter 12
Digestion of Carbohydrates, Proteins and Fats

"Moses also took all the fat around the inner parts, the covering of the liver, both kidneys and their fat, and burned it on the altar"

Leviticus

The narrative above is likely an early reference to visceral fat. Digestion of food begins in our mouth with salivary amylase that acts on carbohydrates. The taste of food is only perceived by gustatory receptors on our tongue. Once swallowed, the enjoyment of any taste is over. So chew more, especially on what is called 'fast food'!

Metaphysiologically, it is a wonderful coincidence that digestive enzymes are secreted from our exocrine glands that hydrolyze the three major types of nutrients in our food, namely, carbohydrates, proteins, fats (and also nucleic acids, DNA, RNA).

Enzymes in saliva, gastric and pancreatic secretions act in the mouth, lumen, aided by the mixing action of the GI motility. The penultimate enzymatic actions for peptides and disaccharides to their final amino acid and monosaccharide components respectively are achieved by luminal membrane bound enzymes of the intestinal enterocytes. For small absorbable peptides, there are also intracellular peptidases that release amino acids to exit the cell by amino acid transporters at the basolateral membrane of the intestinal epithelium.

Some of the undigested food are acted upon by resident colonic bacteria, that appears to have important physio-ecological functions.

For fats, pancreatic lipase, phospholipase, cholesterol esterase act on emulsified dietary fats that are generated by bile salt actions, when bile is released by neuroendocine mechanisms, triggered by presence of fats in the duodenum.

H.M. Cheng (ed.), *Physiology Question-Based Learning*,
DOI 10.1007/978-981-10-0877-1_12

1 Carbohydrate Digestion

Explain the following regarding the digestive breakdown of carbohydrates.

a. Cellulose remains undigested in the diet.
b. The only enzyme that debranches α-limit dextrin is isomaltase.
c. Pancreatic amylase does not generate glucose from digestion.
d. Pancreatic amylase does not completely breaks down starch to disaccharides.
e. Disaccharidases are not secreted by the intestine.

Plant starch, amylopectin is the major source of carbohydrates in most human diet. Cellulose, the major content of dietary fiber has β 1,4 linked glucose polymer. These glycosidic linkages cannot by hydrolyzed by intestinal enzymes. The salivary and pancreatic α-amylase cannot cleave α-1,6 linkages and terminal α 1,4 bonds in the carbohydrate. The products released are maltose, maltotriose and α-limit dextrins. Isomaltase or α-dextrinase is a brush border oligosaccharidase. It breaks down the α-1,6 linkages at the branch points of the α-limit dextrins.

Pancreatic digestion of starch also produces malto oligosaccharides (4–9 glucose units) and α-limit dextrins (5–9 glucose subunits). Glucoamylase is another brush border enzyme that releases one glucose at a time from malto oligosaccharides.

Disaccharidases are also found in the mucosal epithelium of the duodenum and jejunum. Lactase cleaves lactose to glucose and galactose. Sucrase breaks down sucrose into fructose and glucose.

Sucrase and α-dextrinase are noncovalently associated as subunits of a single protein. After insertion unto the brush border membrane, pancreatic proteases hydrolyze it into the two enzyme parts.

Brush border trehalase cleaves trehalose, a 1,1 α-linked disaccharide of glucose. Trehalose is found in mushrooms and yeast.

2 Protein Digestion

Explain the statements below regarding protein digestive processes.

a. Ingested proteins is not the only protein source for digestion.
b. Pepsin is non-essential for protein breakdown.
c. Small peptides are the major digestive products rather than amino acids.
d. The digestion and absorption of all protein sources is almost complete.
e. An intestinal enzyme converts all the trypsinogen to trypsin. Why is this statement incorrect?

Besides ingested protein, proteins are also available in the form of digestive enzymes and exfoliated epithelial cells. A small amount of protein is present only in feces. This is mainly derived from colonic bacteria, exfoliated cells and mucoproteins in colonic secretions. In human, by the end of the jejunum, ingested protein is almost all absorbed. To balance normal catabolism, the required protein/day is about 0.7 g/kg body weight.

Maximally, pepsins break down about 15 % of dietary proteins. The main products of protein digestion by pancreatic proteases and brush border peptidases are small peptides and amino acids. The former is three to four times more than the amino acids. Inside the intestinal cell, cytosolic peptidases release amino acids, especially from dipeptides and tripeptides.

The enzyme, enteropeptidase on the brush border membrane of duodenum and jejunum activates trypsinogen to trypsin. Trypsin then acts autocatalytically to convert more trypsinogen to trypsin. Trypsin also cleaves the other pancreatic proenzymes to their active forms.

The brush border peptidases include amino peptidase and dipeptidyl aminopeptidase.

3 Fat Digestion

a. What is the principal form of phospholipids that is absorbed?
b. What is the diameter of a mixed micelle?
c. What products of fat digestion mainly form micelles with bile acid?
d. How are fat-soluble vitamins absorbed?
e. How does the micelle facilitate the brush border absorption of lipid digestion products?

Phopholipase A_2 cleaves the 2-possition ester bond of a glycophosphatides. If the substrate is phosphatidylcholine, the products will be one free fatty acid and one lysophosphatydylcholine. Prophospholipase A_2 is activated by trypsin. Both phospholipase A_2 and pancreatic lipase do not break well the fatty acyl ester linkage at the 1 position. Thus phospholipids are mainly absorbed as lysophosphatides.

Bile acids generate micelles especially with 2-monoglycerides and lysophosphatides. The hydrophobic acyl chains of 2-MG and lysophosphatides are orientated towards the interior of the micelles. Bile acids are flat amphiphatic molecules that have a polar and a non-polar domains. The non-polar face of the bile acid is directed towards the lipid interior of the micelles. Each micelle is about 5 nm in diameter and contains about 20–30 lipid molecules. The micelles are small enough to diffuse among the microvilli and facilitate the absorption of lipids. Very hydrophobic molecules like long-chain fatty acids, cholesterol and fat-soluble vitamins preferentially partition into the micellar interior. Fat-soluble vitamins, A, D, E, K are poorly absorbed if fat digestion and absorption is produced by pancreatic enzyme deficiency or obstructed bile flow.

Micelles keep the aqueous solution around them saturated with 2-MG, cholesterol, fatty acids and lysophosphatides. Because these lipids are very insoluble in water, their aqueous concentrations are low. The mixed micelles diffuse through the unstirred layer. Thus the aqueous solution in contact with the brush border is saturated with products of fat digestion, ready for absorption over the large surface area of the brush border.

4 Digestion of Lipids

a. Does the stomach digest lipids?
b. Are the lipolytic pancreatic enzyme water-soluble?
c. Are bile acids good emulsifying agents?
d. What is the diameter of an emulsion droplet?
e. Bile salts inactivate pancreatic lipase. How does the intestinal lipid digestion proceed?

A fair amount of triglyceride digestion takes place in the stomach. The enzymes that hydrolyze lipids in the gastric phase are called pre-duodenal lipases. The enzyme activities are optimal at acid pH. Gastric lipase is produced by glands in the fundus of the stomach. If pancreatic lipase is deficient or inactivated by high acidity in the intestine, the contribution of gastric lipase to the total hydrolysis of the triglycerides becomes significant.

Pancreatic lipolytic enzymes have access only to the surface of fat droplets. Emulsification increases the total surface area for lipid digestion many thousand times. Each emulsion droplet is about 1 μm in diameter. Alone, bile acids are weak emulsifiers. In combination with the phospholipid lecithin, available in high concentration in the bile, dietary lipids are effectively emulsified. Pancreatic lipase (glycerol ester hydrolase) acts at the interface between the aqueous phase and the triglyceride oil phase. Bile salts bind to the oil droplets and prevent the lipase from acting. A cofactor in pancreatic juice, colipase is able to displace bile salts from the oil droplets. The colipase forms a complex with the pancreatic lipase. Together the enzyme complex cleaves the 1 and 1′ fatty acids from a triglyceride. This releases two free fatty acids and a 2-monoglyceride.

5 Bile

a. Where are the two major organic constituents of bile?
b. How many types of bile salts are present in bile?
c. Why are bile salts more soluble than bile acids?
d. Why does a pancreatectomized animal have steatorrhoea?
e. Why does ileal resection lead to steatorrhoea?

Bile acids (50 %) and phospholipids (40 %) are the primary organic compounds of bile. Bilirubin (2 %) and cholesterol (4 %) are also present. The total bile acid pool is about 2.5 g (includes bile acids in the liver, gallbladder, bile ducts and intestines.

There are two primary bile acids, cholic acid and chenodeoxycholic acid. Intestinal bacteria dehydroxylate them into secondary bile acids, deoxycholic acid and lithocholic acid.

The hepatocytes conjugate bile acids with glycine or taurine. There are thus eight bile salt types, named after the parent bile acid and the conjugated amino acid (e.g taurocholic acid).

The pKs of bile acids are about 7.0. At the duodenal pH of pH 3 to 5, bile acid will be mostly nonionized and water-insoluble. Bile salts have pKs between 1 and 4. In the duodenal luminal environment, bile salts will be considerably ionized and soluble in water.

If the exocrine function of the pancreas is absent, steatorrhoea occurs because of lack of pancreatic lipase. The absence of pancreatic bicarbonate also accounts to the fatty stools. Acid not only inhibits pancreatic lipase but may also precipitate some bile salts.

In a patient after ileectomy, the fecal loss of bile acids is great and new synthesis of bile acids is stimulated. However, the hepatic production cannot keep pace with the excreted loss. The bile acid pool is reduced. Emulsification and micellar formation become inadequate. Excess dietary lipids are excreted with the feces.

High concentrations of unreabsorbed bile acids in the colon stimulate chloride secretion by the crypt cells. This leads to a secretory diarrhea.

6 Bilirubin Metabolism

a. What is the major source of bilirubin?
b. Is bilirubin protein bound inside the hepatocyte?
c. Is there recycling of bilirubin from the intestine?
d. When is jaundice observable?
e. When does conjugated bilirubin levels become elevated in blood?

Age erythrocytes are taken up by the tissue macrophages. The heme portion is converted to bilirubin by heme oxygenase. In humans, most of the biliverdin is changed to bilirubin. Bilirubin in the circulation is albumin-bound. Upon entering the hepatocytes, the bilirubin becomes associated with cytoplasmic proteins.

Bilirubin is then conjugated to glucuronic acid in a 1:2 ratio in the smooth endoplasmic reticulum. The glucuronide is more water soluble and is actively transferred into the bile canaliculi. Some conjugated bilirubin leaks back into the blood and is excreted in the urine. Therefore total plasma bilirubin comprises free bilirubin and a small quantity of bilirubin glucuronide.

The intestinal mucosa is permeable to unconjugated bilirubin and urobilinogens, formed by intestinal bacteria. When free or conjugated bilirubin is retained in the blood, jaundice develops. Jaundice is seen when the plasma bilirubin exceeds 2 mg/100 ml (or 34 μmol/L). Hyperbilirubinemia can be associated with predominantly increased free or conjugated bilirubin. The former case is linked with hemolytic anemia, decreased hepatic uptake or defective glucuronide can regurgitate back into the circulation when there is interrupted secretion of conjugated bilirubin into the bile canaliculi. Intra-or extrahepatic bile duct obstruction also increases the plasma concentration of conjugated bilirubin.

7 Bile Flow

Comment on the following regarding bile secretions.

a. Drugs that block bile acids reabsorption lower blood cholesterol.
b. Bile is concentrated by the bile bladder epithelium.
c. The bile acid pool is sufficient for a typical meal.
d. The highest rate of gallbladder emptying takes place during the intestinal phase of digestion.
e. The bile duct epithelium secrets an aqueous secretion.

Bile acids → excreted in the feces is the only significant pathway of cholesterol removal. Bile acids are synthesized by the hepatocytes from cholesterol. Thus agents that block the ileal reabsorption of bile acids will activate new synthesis of bile acids from cholesterol.

Between meals, bile is diverted into the gallbladder. The gallbladder concentrates the bile acid concentration 5 to 20-fold. Active transport of sodium is the mail active process in the concentrating action. The epithelia of the gallbladder have tight junctions and water absorption occurs by the standing osmotic gradient mechanism.

The strongest stimulus for gallbladder contraction is cholecystokinin. CCK relaxes also the sphincter of Oddi, which guards the entrance of the common bile duct into the duodenum. Normally, the rate of gallbladder evacuation is sufficient to keep the concentration of bile acids in the duodenum above the critical micelle concentration.

Bile acids emulsify fats to increase the surface area for digestion by pancreatic lipolytic enzymes. Bile acids then formed mixed micelles with the lipid digestion products. The transport of these products by the micelles to the epithelial brush border promotes the lipid absorption. Enterohepatic circulation occurs two to more times with a meal. About 10–20 % of bile acids escape absorption and is lost in the feces.

The aqueous secretion of the bile duct is isotonic and contributes about 50 % of the bile volume. The Na^+ and K^+ concentrations are similar to that in plasma. The HCO_3^- content is high and the hormone secretion stimulates this function of the bile duct.

8 Micelles and Lipid Digestion

a. Of the classes of nutrients, which types are more prone to malabsorption?
b. How much of intact triglycerides is found in micelles?

c. Does the enterocyte make chylomicrons in the absence of luminal fat?
d. Can triglyceride be hydrolyzed in the absence of bile acids?
e. In the absence of pancreatic lipases, why are all classes of lipids poorly absorbed?

The digestion and absorption of lipids are more complex than for other classes of food. Thus malfunction of lipid absorption is more frequently accounted.

Mixed micelles are formed by lipid digested products and bile salts. The two major constituents are lysophosphatides and α-monoglycerides. In the hydrophobic intense of the micelle, extremely non-polar molecules like fat-soluble vitamins, cholesterol and long chain fatty acids tend to partition. Hardly any undigested triglyceride is sequestrated in the micelle.

Chylomicrons are not formed in the absence of fats in the diet. However, the enterocytes do make very-low density lipoproteins (VLDL) and extrude them into the intestinal lymph. VLDL has less triglycerides (60 % of VLDL mass) and more protein. They are more dense than chylomicrons.

Even when bile acids are not available, up to 50 % the normal fatty acid absorption from triglycerides can occur. Cholesterol and fat-soluble vitamins absorption are more severely affected by the absence of bile acids. Pancreatic insufficiency is also a cause of lipid malabsorption. This is probably due to the lack of building blocks for assembling the lipid micellar carrier. Lysophosphatides and 2-MG are both products of pancreatic lipase enzymatic activity.

9 Intestinal Epithelial Handling of Lipids

a. How is fatty acids transported in the cytosol of the enterocyte?
b. Why is chylomicron absorbed into the lacteals instead of the portal circulation?
c. What contributes mainly to the mass of chylomicrons?
d. Why does chylomicron vary so much in size?
e. What molecule makes up most of the chylomicron surface?

Lipid resynthesis occurs in the intestinal epithelial cells before they are absorbed. This takes place in the smooth endoplasmic reticulum. The SER becomes engorged with lipids after a meal. Fatty acid-binding proteins (FABP) function to transport products of lipid digestion from the brush border to the SER. One type of FABP binds long-chain fatty acids. The other type of FABP has a broader specificity and carries also cholesterol, monoglycerides and lyophosphatides. There are also sterol carrier proteins. The re-esterification of 2-MG and lysophospholipids is virtually complete. Fatty acids with less than 10–12 carbon atoms can enter the portal blood as free, unesterified fatty acids.

The prechylomicrons from the SER are moved to the Golgi apparatus and further processed to chylomicrons. The chylomicrons are extruded by exocytosis. The remaining 20 % are taken up by the apolipoproteins. Most of the weight of the chylomicrons (90 %) comprise triglycerides which occupy the core of the chylomicron peptide. Cholesterol and cholesterol esters are found together with the TG and represents a small 1 % of chylomicron mass.

The size of the chylomicron depends on the amount of lipids that are absorbed. Chylomicron sizes vary from 60 to 750 nm in diameter. Chylomicrons are too large to transverse the basement membrane of the mucosal capillaries. In the lacteals, there are large fenestrations that the chylomicrons can enter through. Chylomicrons are drained by the lymphatic flow via the thoracic duct into the venous circulation.

10 Bile Acid Absorption and Secretion

a. Is the intestinal absorption active or passive?
b. Does bile contain primary or secondary bile acids?
c. What is the effect of bile acids in the portal blood on hypatocyte bile acid synthesis?
d. What are the constituents of gallstones?
e. Bile acids are secreted into canaliculi down its concentration gradient. Comment.

Conjugated bile acids are actively absorbed at the terminal ileum by a Na^+-bile salt cotransport basolateral Na^+/K^+ ATPase system. Bacteria in the terminal ileum and colon deconjugate bile acids and also dehydroxylate them to secondary bile acids. These two reactions reduce the polarity of bile acids and they can also be absorbed by simple diffusion. The hepatocytes reconjugate the bile acids normally with glycine or taurine. Rehydroxylation also occurs. The uptake and resecretion of bile acids by hepatocytes is stimulated by the reabsorbed bile acids in the enterohepatic cycling. This choleretic effect is associated with inhibition of new bile acid synthesis. Ileal resection increases the de novo synthesis of bile acids.

If the bile is supersaturated with cholesterol, the excess that is not solubilized in the micelles tend to crystallize. The most common variety of gallstone is cholesterol gallstones. Bile pigment gallstones can also form. This is mainly the insoluble calcium salts of unconjugated bilirubin. In liver disease, hepatocytes may be poor in making glucuronides of bilirubin.

Bile acids are secreted probably by facilitated diffusion. The concentration gradient is partly maintained by micellar formation in the canaliculi. This keeps the free concentration in solution quite low (critical micelle concentration).

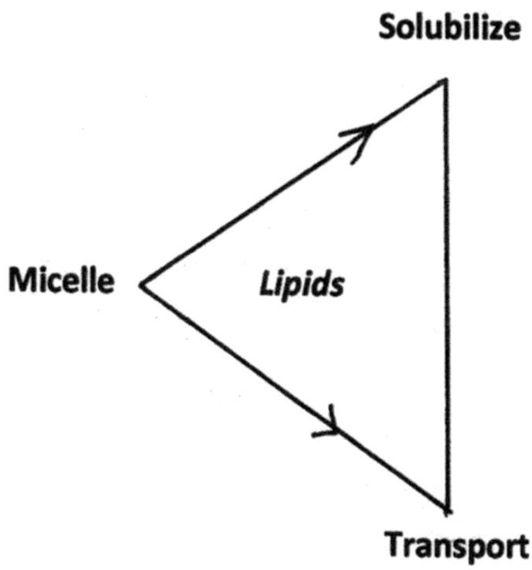

The digestion and absorption of fats in the watery medium of the intestinal lumen requires the essential role of bile salts secreted in gall bladder bile when it contracts during a meal as fatty chyme enters from the stomach. The bile salts form mixed micelles which serve to solubilize the products of lipolysis and also to traffic them in these micelles to the absorptive epithelium. Micelles are not absolutely required for assimilation of lipids as fatty acids and monoglycerides have sufficient aqueous solubility to diffuse across the epithelium. However, fat soluble vitamins and cholesterol can't diffuse by 'themselves and need micelles'!

The release of duodenal secretin or cholecystokinin (CCK) is physio-logically regulated by a homeostatic feedback loop. For secretin, this is acid > Secretin > bicarbonate >neutralize acid. For CCK, the circular feedback sequence is protein/lipids > CCK > protease/lipase > proteo/lipolysis.

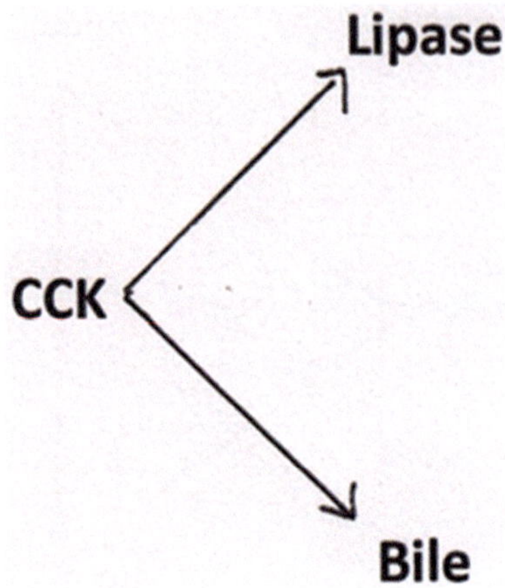

The hormone cholecystokinin (CCK) has a dual action on the gallbladder and pancreatic acinar cells. Bile and an enzyme-rich pancreatic secretion are released by CCK respectively. The bile salts in a fatty collaboration with pancreatic lipase act and digest dietary triglycerides.

Billy is a responsiBile character who lives in the green Gall Uplands. He comes down to the Duodenum Valley occasionally and helps to transport the fat people there.

Chapter 13
Intestinal Absorption

In the human body, epithelium serve common transport mechanisms. Epithelial cells are functionally polarized as determined by the different set of membrane transporters on the luminal and basolateral side of the cell. This allows unidirectional absorption or secretion process; absorption going from lumen across the luminal and basolateral membranes into the interstitial space and hence into the capillaries. Secretion will be from the capillaries, trans-epithelially into the lumen.

Some electrolytes and water can also move paracellularly, if the inter-cellular tight junctions do not occlude their movement.

The renal epithelium of the nephron shares several matched transport mechanism with the intestinal enterocyte. Glucose is absorbed trans-epithelially by identical secondary active transport in both the intestines and at the nephrons. There are more established renal tubular secretory transport events compared with the GI, since the kidneys provide a major excretory route for endogenous and exogenous organic metabolites.

Incidently, the epithelium of the ducts in both the salivary and pancreatic ducts modify the primary acinar secretions by absorption /secretion to produce the final fluid in the mouth and duodenum respectively.

The intestines has a major absorptive function for all three energy substrates, monosaccharides, amino acids and fatty acids. Water movement in both the intestinal and renal epithelium follows solute absorption which generates a local osmotic gradient that drives the water across the epithelium. Dysruption of this osmotic gradient occurs in malabsorption of lactose (intolerance with osmotic diarrhea) and glucosuria (osmotic diuresis).

Bile salts are also reabsorbed by a sodium linked secondary active mechanism . The bile salts enter the enterohepatic circulation to recycle sufficient bile salts for optimal intestinal lipolysis and assimilation. Most of the lipids are absorbed in a chylomicron structure that enters the lymph vessels. Our hearts are privileged to be the first organ to meet these postprandial fatty lymphatic streaks as they reenter the systemic circulation.

H.M. Cheng (ed.), *Physiology Question-Based Learning*,
DOI 10.1007/978-981-10-0877-1_13

1 Absorption of Carbohydrate

Correct the following inaccuracy concerning intestinal transport of carbohydrate end products.

a. Fructose, galactose and glucose cross the basolateral membrane of the intestinal epithelial cell in the same way.
b. The glycemic index of all starch-containing food is the same.
c. Lactose intolerance is due to deficiency of membrane receptor for lactose absorption.
d. In oral sugar tolerance test, the sugar will appear in both the feces and blood in the intolerance subject.
e. All malabsorption conditions are due to low activity of oligosaccharidases in the intestinal cell membrane.

Fructose is not a substrate for the brush border glucose-galactose transporter (AGLT; sodium-glucose transport protein). Fructose enters by facilitated diffusion on GLUT 5. All the three monosaccharides then leave the cell by facilitated diffusion across the basolateral membrane by the GLUT 2 transporter. Dietary starch is normally not completely absorbed. The rate, extent of digestion and absorption of various starch meals is expressed as a glycemic index. In this measurement, a 50 g carbohydrate containing food is given and the glucose in blood is monitored for 2 h. The percentage of increased glucose detected in blood relative to that after consuming 50 g of glucose is the glycemix index. For e.g. carrots have a high index of 90 % and that of apples is 40 %.

Lactose malabsorption is caused by deficiency of intestinal lactase. The undigested lactose is actively metabolized by colonic bacteria. They release gases and metabolites that increase colonic motility. More than 50 % of adults in the world are lactose intolerant. In any carbohydrate malabsorption syndrome, the administrated sugar will all end up in the feces during an oral sugar tolerance test. Very rarely, a mutation in the SGLT, the glucose and galactose transporter occurs. Consumption of glucose, galactose or starch will result in flatulence and severe diarrhoea. Fructose is still well tolerated and can be given to infants with this condition.

2 Absorption of Protein Digestion Products

Elaborate on the accuracy of the following descriptions of intestinal transport of protein products.

a. Peptides are carried into intestinal cells by sodium cotransport mechanism.
b. Amino acid transporters at the brush border are sodium dependent.
c. Basolateral membrane also has transporter for amino acids to move into the blood.
d. Intact proteins are not absorbed.
e. Deficient amino acid absorption will result in protein malnutrition.

Small amount of luminal proteins are engulfed by M cells of the mucosal immune system. In ruminants and rodents (not in human), the intestine of the young endocytose immune proteins in colostrums by receptor mediated mechanism.

A single type of secondary active transport absorbs small peptides. This system especially transports di and tripeptides and it is stereospecific for the physiological L-amino acids. The transfer across the membrane is driven by an electrochemical gradient of H^+. A Na^+/H^+ exchanger present also on the luminal membrane generates an acidic microenvironment near the brush border surface. The Na^+/H^+ exchanger is in turn powered by the basolateral Na^+/K^+ ATPase.

There are five transporter types for amino acids for neutral, basic, acidic and imino acids (proline and hydroxyproline). Three of these are Na^+-dependent and include the neutral and acidic amino acids. In the basolateral membrane two Na^+-linked cotransporter brings amino acids into the epithelial cells to provide for protein synthesis in-between meals. The other three Na^+-independent pathways are responsible for the efflux of the amino acids into the blood. For acidic amino acids (e.g. glutamate, aspartate), there are no established basolateral transporters. These amino acids once absorbed across the luminal side are presumably shunted to biochemical reactions of energy metabolism within the cells. Defects in amino acid absorption (e.g. Hartnup's disease, cystinuria or prolinuria) does not necessary lead to protein lack as the epithelial cells can still absorb the dipeptides and tripeptides.

3 Absorption of Lipids

a. How much of ingested fats is normally absorbed?
b. What is the limiting step in the absorption of lipids?
c. How wide is the unstirred layer at the brush border?
d. Why is there a lipid concentration gradient across the unstirred layer?
e. Are plant sterols more easily absorbed compared to cholesterol from animal sources?

In normal stools, the fat is not derived from undigested fat in the diet. Fat consumed is normally completely absorbed. Fecal fat is colonic bacterial in origin as well as lipids from exfoliated intestinal epithelial cells. The products of lipid digestion diffuse rapidly at the brush border plasma membrane. The major limiting step in lipid absorption is the diffusion of the mixed micelles through the diffusion boundary layer (unstirred layer) on the luminal surface of the brush border. This unstirred layer is 200–500 μm in thickness. Since the intestinal mucosal surface is convoluted, the fluid in close contact with the epithelial cells are not well mixed with the bulk of the GI luminal contents. There is thus a concentration gradient across the unstirred layer. The micelles and lipid digestion products are higher in concentration at the luminal side of the unstirred layer than the brush border mucosal side.

The FFA, 2-MG, lysophosphatides and cholesterol easily diffuse across the brush border membrane. These are however transport proteins on the luminal membrane that transfer long-chain fatty acids.

The microvillus membrane fatty acid-binding protein (MVM-FABP) takes up the FFA into the enterocytes in a Na^+-linked, secondary active mechanism.

About 50 % of ingested cholesterol is absorbed. The slower absorption of cholesterol means that as the micelles move down the small intestine, they become progressively enriched with cholesterol. For plant sterols, only about 2 % of the amount consumed is absorbed. This is due to an ATP-Binding Cassette (ABC) transporter that pumps cholesterol back into the intestinal lumen.

4 Regulation of Intestinal Secreto-Motor Functions

Specify the following regarding the neuro-hormonal mechanisms of intestinal functions.

a. Most of the direct innervation of intestinal epithelial cells comes from the enteric neurons.
b. Reflexes in the enteric nervous system influence secretomotor activity of the intestine.
c. Parasympathetic fibers modulate intestinal absorption and secretion.
d. Somatostatin inhibits secretion.
e. GI mast cells release modulators that stimulate secretion.

The epithelial cells of the intestine receive innervation mainly from the submucosal ganglia. The major influences of the parasympathetic and sympathetic neurons are effected via their action in the enteric neuronal functions. Parasympathetic input reduces absorptive fluxes and stimulates secretion. Basal rate of secretion may be due to parasympathetic activity. Conversely, sympathetic discharge enhances net absorption. The autonomic neuropathy in diabetics could contribute to diabetic diarrhoea due to the lower sympathetic outflow.

Various stimuli, distention, food constituents, bile salts, pH, antigen, etc activate enteric neuronal reflexes. All these stimuli produce reflex stimulation of secretion. Definite enteric neurons that inhibit intestinal epithelial secretion have not been identified.

Somatostatin decreases the secretion by crypt cells. Somatostatin analogs have been used to treat secretory diarrhoea. Somatostatin is co-released with noradrenaline at some sympathetic nerve terminals. This neurotransmitter may also cause enteric neurons to reduce the secretomotor outflow to intestinal enterocytes.

Substances released from primed mast cells and other GI immunocytes elicit secretion in two ways. The mediators directly stimulate secretion and/or inhibit absorption. The cellular products also act on enteric neurons to increase the secretomotor activity. The enteric neurons in turn can also affect the function of the GI mast cells and immunocytes.

5 Absorption of Water

Elaborate on the following statements regarding intestinal movement of water.

a. Most of the water absorbed is from gastrointestinal secretions.
b. In the duodenum, net movement of water is from blood to lumen.
c. The largest amount of water is absorbed by the small intestine.
d. Water absorption in the small intestine occurs in the absence of a transmucosal osmotic gradient.
e. Does intestinal cells also secrete water?

On average, about 2 L of water is consumed/day. The water source from gastrointestinal secretions is much larger, about 7 L/day. The GI secretion per day roughly consist of 1.5 L saliva, 2 L gastric juice, 500 ml bile, 1.5 L pancreatic juice and 1.5 L intestinal secretions. The greater amount of water absorption occurs in the small intestine. Approximately 8.5 L/day is absorbed, more in the jejunum than in the ileum. In the duodenum, the gastric chyme is frequently hypertonic. Further digestion increases the osmotic components even more. Thus the net flux of water in the duodenum is in the direction of blood to lumen to isotonize the chyme.

Of the measuring 500 ml, the colon absorbs 400 ml/day. Only about 100 ml/day of water is excreted in the feces.

Water reabsorption in the small intestine takes place in the absence of an osmotic pressure difference between the luminal contents and the blood in the intestinal capillaries. The fluid at the apical end of the intercellular space is hypertonic. As a result of osmotic flow from adjacent cells, the fluid becomes less hypertonic as it flows down the intercellular channel. At the serosal or basal end of the channel, the fluid is virtually isotonic.

The highly, differentiated mature epithelial cells at the villus tip are specialized for absorption of ions and water. The immature, less differentiated cells in the crypts of Lieberkuhn normally function to produce a net secretion of electrolytes and water.

6 Pathophysiology of Salt and Water Absorption

a. What happens if the Cl^--HCO_3^- exchanger in the brush border of the colon is non-functional?
b. What contributes to osmotic diarrhoea?
c. How is intestinal hypermotility associated with fat malabsorption?
d. How does cholera toxin induce hypersecretion of water and electrocytes?
e. How do opioids act as antidiarrhoeal agent?

In congenital chloride diarrhoea, the luminal Cl^--HCO_3^- exchanger is defective. The Cl^- concentration exceeds the combined concentrations of Na^+ and K^+. Normally, a parallel Na^+/H^+ exchanger accounts for the NaCl absorption by the colon. H^+ continues to be secreted without the HCO_3^- to neutralize it. The net loss of H^+ contributes to a metabolic alkalosis.

Abnormal absorption of a non-electrolyte (nutrient) results in an osmotic diarrhoea. This occurs in any of the malabsorption syndromes involving carbohydrates. Water is retained in an increased volume of chyme. This overwhelms the colonic ability to absorb the water and electrolytes. Malabsorption also occurs in conditions that result in a decreased surface area for absorption. In celiac disease (or celiac sprue), the height of the villi is reduced and the number of microvilli is also less. This disorder is due to an immune hyperreactivity to gliadin in wheat gluten. In fat malabsorption, the colonic bacteria metabolize the lipids and hydroxyl fatty acids are released. These metabolites stimulate colonic motility and also inhibit electrolyte and water absorption by the colonic epithelium.

Cholera toxin elevates the cAMP in the Lieberkuhn's crypt cells. The unidirectional secretory flux exceeds the unidirectional absorptive flux by other cells in the villi. A secretory diarrhoea results. In pancreatic cholera, elevated VIP produces the same effect.

Opioids act on δ-receptors to stimulate electrolyte and water absorption. Opioids also suppress intestinal motility. The combined actions contribute to their effects in alleviating diarrhoea.

7 Absorption of Vitamin B_{12}

a. Will malabsorption of vitamins B_{12} result in symptoms in a few weeks?
b. Why does vitamin B_{12} deficiency occur in pancreatic insufficiency?
c. Is vitamin B_{12} absorption powered by Na^+ in a secondary active transport mechanism?
d. Why does the colon not absorb vitamin B_{12} although enteric bacteria can synthesize the vitamin?
e. Can pernicious anemia be treated with vitamin B_{12} alone?

Some of the water-soluble vitamins are absorbed by a Na^+-linked active transport system. This includes vitamin C, pantothenic acid and thiamin (B_1). For vitamin B_{12}, the absorption mechanism is receptor-mediated endocytosis. The membrane receptors for B_{12} absorption is localized on the ileal epithelial cell membrane. The ileal receptors only bind to vitamin B_{12} when it is complexed with intrinsic factor that is secreted from the gastric parietal cells. Normally, vitamin B_{12} binds strongly to other proteins in gastric juice, called R proteins. Pancreatic proteases are required to digest and reduce the B_{12}-R binding affinity so that vitamin B_{12} is available to interact with IF, its absorptive co-factor.

The vitamin B_{12}, absorbed into the portal blood is transported bound to transcobalamin. The hepatocytes take up the transcobalamin-B_{12} complex by receptor-mediated endocytosis. The liver has a large store of vitamin B_{12} (about 5 mg). Most of the vitamin B_{12} present in bile is reabsorbed. Only about 0.1 % of the hepatic store is lost daily. In the total absence of vitamin B_{12} absorption, the store can still supply the body's need for about 5 years.

In very large doses (1 mg/day), about 2 % of the oral load of vitamin B_{12} can be absorbed. This IF-independent mechanism is not restricted to the ileum and is not saturable.

8 Absorption of Iron

a. Why is iron absorption limited?
b. Is heme iron digested to release iron for absorption?
c. How does iron move across the luminal brush border?
d. In what form is iron absorbed into the portal blood?
e. How is iron carried in the blood?

In a typical Western adult diet of 15–20 mg/day, only about 1 mg/day of iron is absorbed by normal males. It is slightly higher at around 1–5 mg/day. Absorption of iron is limited because iron easily forms insoluble salts with anions in intestinal secretions as well as common food constituents like tannin, phytate and fiber. Gastric HCl does increase the solubility of these iron complexes. The tendency of Fe^{2+} to form insoluble complexes is much less than for Fe^{3+}. The antioxidant, vitamin C reduces Fe^{3+} to Fe^{2+}. Ascorbic acids also forms a soluble complex with iron.

Heme is released from proteins by digestive proteases. The heme is absorbed by a heme transporter. Inside the enterocytes, iron is liberated from the heme by heme oxygenase. This enzymatic reaction is the rate-limiting step in the absorption of iron.

Inorganic or non-heme iron is cotransportetd into the epithelial cells by a Fe^{2+}/H^{+} symporter. This transporter does not carry ferric iron. A brush border iron reductase is available to aid in the absorption. In the cytosol, the ferrous iron is oxidized by a ferroxidase. Fe^{3+} is bound to cytosolic iron-binding proteins. This reduces the possibility of Fe^{3+} to form insoluble complexes with intracellular anions. Fe^{3+} crosses the basolateral membrane by another iron transporter. In the circulation, ferric iron is bound to the protein, transferrin. Cells that utilize iron have membrane receptors for the transferrin-Fe^{3+} complex, which then enters the cells by receptor-mediated endocytosis.

9 Regulation of Iron Absorption

a. How many atoms of iron is associated with ferritin?
b. What is the role of ferritin in iron absorption and homeostasis?
c. What is the relationship between transferring-iron and iron regulatory protein in the enterocytes?

d. What is the fate of ferritin?
e. How soon after hemorrhage does intestinal iron absorption increase?

Apoferritin is the protein part of ferritin. Ferritin is the storage form of iron in the enterocytes. Ferritin is a large complex of 24 apoferritin; holding up to 4000 iron atoms. Iron bound irreversibly to ferritin is no longer available for transport into the blood. Thus the regulatory effects on synthesis of iron transporters (at the luminal and basolateral membranes) and apoferritin are reciprocal. In iron deficiency, there will be increased translation of mRNAs for iron transporters and reduced translation of apoferritin mRNA. This reciprocal action is carried out by a cytosolic iron regulatory protein in the enterocytes. The amount of apoferritin will affect how much iron is entrapped in this non-absorbable pool.

Apoferritin-bound iron is lost into the intestinal lumen and excreted in the feces when the intestinal epithelium exfoliates. About 3–4 days after blood loss, the duodenal and jejenal absorption of iron is elevated. During this time, the absorptive cells migrate from the crypts of Lieberkuhn to the tips of villi. The crypt cells have basolateral receptors for circulating transferring-iron (Trf-Fe). When circulating Trf-Fe is high, uptake increases the cytosolic Fe^{3+}. Binding of Fe^{3+} to iron-regulatory protein (IRP) decreases the synthesis of the iron transporters and stimulates apoferritin production.

The iron absorptive ability of the enterocytes is thus determined or "programmed" when the cells are in the crypts of Lieberkuhn before the functional migration upwards towards the absorptive surface of the villus.

10 Absorption of Calcium Cations

Elaborate on the following physiology of intestinal calcium transport

a. The cytosol of intestinal epithelial cells contains calbindins.
b. Calcium leaves the enterocyte by primary and secondary active transport.
c. Gastric acid secretion aids calcium absorption.
d. Parathyroid hormone stimulates intestinal Ca^{++} absorption.
e. Some calcium is absorbed, involving exocytosis.

Calcium tends to form insoluble salts with several anions present in foods including oxalate, phosphate and phytate. The salts are soluble in acidic pH medium. So gastric HCl facilitates the calcium absorption. Calcium enters the intestinal epithelial cells via calcium channels down its electrochemical gradient. In the cytosol, Ca^{++} is bound by calbindins, These proteins allow a great amount of Ca^{++} to move across the cytosol. The intracellular free Ca^{++} concentration is kept low to prevent the formation and precipitation of insoluble Ca^{++} salts.

At the basolateral membrane, there is a Ca^{++} ATPase and a Ca^{++}/Na^{+} exchanger. The Na^{+}/Ca^{++} is more effective when the cytoplasmic Ca^{++} level is high. When intracellular Ca^{++} concentration is low, the Ca^{++} ATPase is the principle pathway for the efflux of Ca^{++}.

The capacity of the intestine to absorb Ca^{++} is regulated. Hypocalcemia activates secretion of parathyroid hormone (PTH). PTH increases the absorption of Ca^{++} through its action in stimulating renal synthesis of active vitamin D. Vitamin D stimulates each phase of intestinal Ca^{++} absorption. Vitamin D promotes the synthesis of both calbindins and the Ca^{++} ATPase. The calbindin cytosolic level correlates well with the intestinal capacity to absorb Ca^{++}.

Ca^{++} can also be transported transcellularly in membrane vesicles. The vesicular Ca^{++} is extruded at the basolateral membrane by exocytosis. Calbindins also enhances the vesicular route.

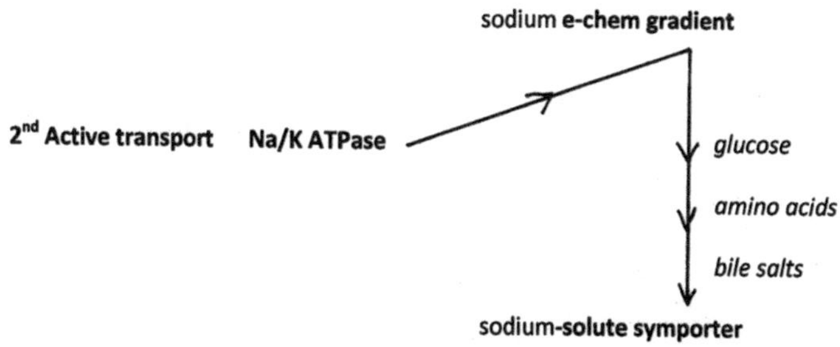

The secondary active transport of solutes that are coupled to sodium electrochemical gradient at the luminal membrane of transporting epithelium. The Na/K ATPase maintains this sodium gradient that carries the solutes separately via their specific symporters into the enterocytes during intestinal absorption.

What is the meaning of 'secondary 'in secondary transport? The playground Slide is a useful analogy. Energy is used indirectly or 'secondarily' in transporting solute by using a potential energy gradient. In most cases this gradient is the trans-membrane sodium concentration gradient, generated actively by the Na/K ATPase. Sodium moves down its gradient and is coupled or linked with the co- transport of another solute (e.g., glucose) by a membrane transporter (symporter).

Imagine the sodium is the mother and the sweet baby is the glucose …and the slide then represents the sodium trans-membrane gradient. Picture credit: Annabella Diong, from the 2016 graduating batch of the Universiti Malaya Medical School, Kuala Lumpur, Malaysia.

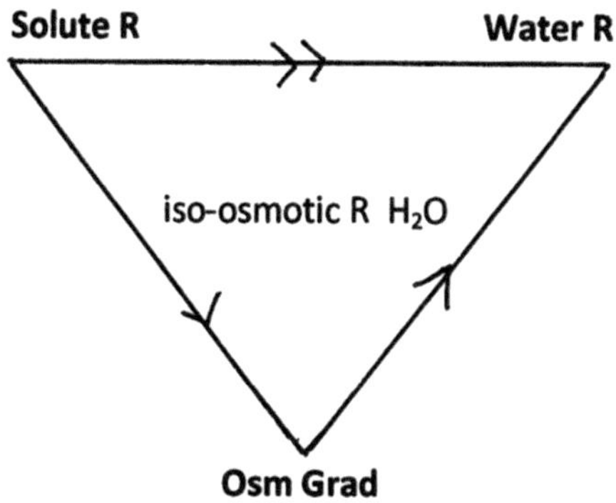

Iso-osmotic reabsorpton of water takes place in both the intestinal and renal tubular epithelium. Water is reabsorbed by an osmotic gradient at the local vicinity, between the luminal fluid and the intracellular fluid or interstitial fluid beyond the paracellular tight junctions. Solute and electrolyte absorption lowers the luminal osmolarity and water movement follows.

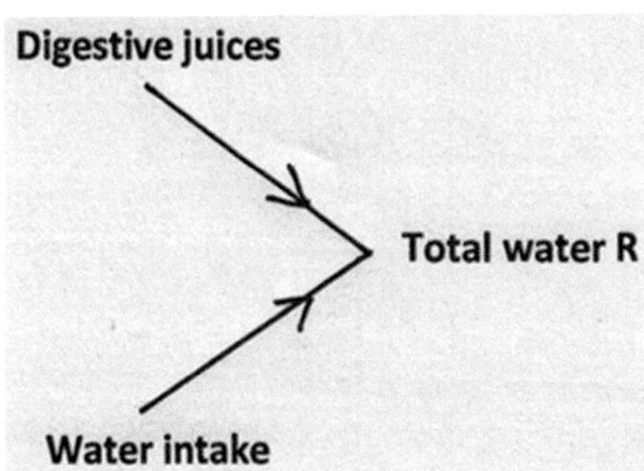

The total volume of digestive secretions (salivary, gastric, pancreatic, bile) exceed the daily water intake by several folds. The water in the juices are recycled or reabsorbed in the small and large intestines. Excessive loss of water in stools during diarrhea can then lead to a negative water balance.

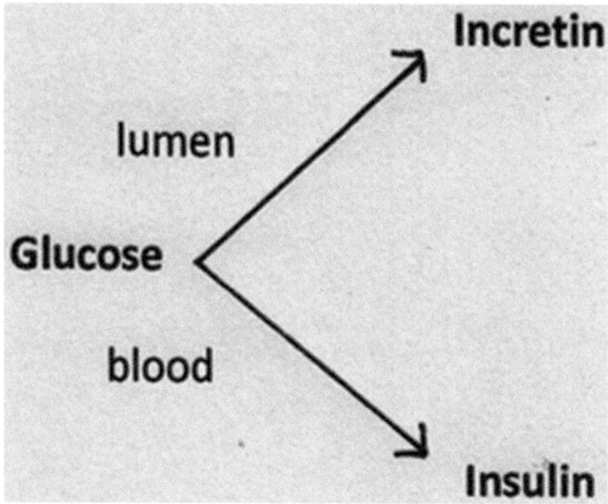

Postprandial rise in glucose is rapidly reversed by pancreatic insulin response. The presence of glucose in intestinal lumen, before absorption causes an 'advance' rise in blood insulin even before blood glucose is elevated. This feed-forward signal is mediated by a family of intestinal hormones termed incretins.

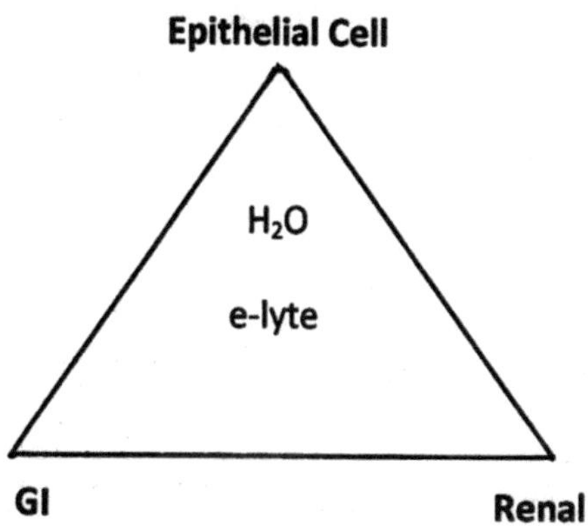

The gastrointestinal (GI) tract and the nephron share the same functional histology of transporting epithelium. The luminal and basolateral membranes at both the intestinal enterocyte and the renal tubular cell have a different spectrum of membrane transporters. This polarized nature of GI and renal epithelium achieves unidirectional absorption or secretion of electrolytes and water.

Chapter 14
Energy Balance, Metabolism

"For life is more than food..." Dr Luke

The blueprint in each living cell is encoded in the chromosomes and DNA. One primary purpose of the cell machinery is to generate energy for its own use via the interconnected biochemical metabolic pathways. Energy, mostly as ATP is needed for work and movement in the multicellular body, tissue repair and maintenance of core body temperature.

The body homeostatically thermoregulates to ensure an optimal temperature for cellular chemical reactions. Does the human body also have sensors for energy balance, for 'fatness'? Are our energy stores as glycogen and triglycerides monitored and used in response to fluctuations around a healthy normal body weight relative to age and gender?

The primary sources of energy substrates is the triad of dietary carbohydrates, proteins and fats. Once digested and absorbed, glycogenesis, lipogenesis and protein synthesis is increased during positive energy balance. There is some biochemical interconversion between glucose, amino acids and fatty acids.

Obesity and Eating disorders in humans highlight the mutifactorial aspects of energy balance that include cultural and behavorial influences. The hypothalamus has a neuronal center that functions like a satiety/hunger meter. Neuro-hormonal mechanisms that are part of the bidirectional brain-gut functional axis affect appetite and anorexia.

The energy scale moves towards negative balance during the inter-digestive period (a more pronounced event during the Muslim Ramadan) and physical activity. Both autonomic neural and particular hyperglycemic and lipolytic hormones are elevated during such energy-lack times to maintain adequate 'bloody fuel' (bloody used here as adjective not as swear word!), especially for cerebral functions.

H.M. Cheng (ed.), *Physiology Question-Based Learning*,
DOI 10.1007/978-981-10-0877-1_14

1 Energy Stores

a. What percentage of total energy stores does carbohydrate contribute?
b. Where is the main glycogen store in the body?
c. What is the dietary sugar necessary for glycogen synthesis?
d. What is meant by diet-induced thermogenesis?
e. What does is mean if the RQ is greater than 1.0?

Carbohydrates contribute 4 Kcal per gram of the substrate. The storage form is glycogen which is a polymer of glucose. Galactose and fructose can also be used as starting materials in glycogenesis.

The stored energy as glycogen represents less than 1 % of the total energy reserves. Three fourths of the glycogen stores in the muscle and the remaining quarter is liver glycogen. Liver glycogen can be broken down to supply energy substrates to other tissues via glycogenesis. Muscle glycogen however is only utilized by the muscle as the tissue lacks the enzyme glucose-6-phosphatase. The reactions of energy storage and energy transfer themselves expend energy. O_2 consumption is stimulated after a meal (diet-induced thermogenesis). Storing glucose as glycogen incurs about 7 % of the original calories. Conversion of carbohydrate to fat uses up much more (>20 %) of the original calories.

When dietary glucose is in abundance, fat metabolism is shifted from fatty acid oxidation to triglyceride synthesis. Some glucose carbon is also converted to fatty acid carbon with a stoichiometry of 11 CO_2/4 O_2 for the reaction from glucose to palmitic acid. The RQ is thus 11/4 = 2.75 which is characteristic when lipogenesis from glucose is taking place.

Inter-organ transfer of energy is also part of whole body metabolism. For example, fatty acid from adipose tissue to the liver and muscle lactate transported to the liver for hepatic gluconeogenesis.

2 Carbohydrate Metabolism

a. What is the proportion of hepatic glucogenolysis and gluconeogenesis that contribute to blood glucose?
b. What is the main substrate for hepatic gluconeogenesis?
c. Can gluconeogenesis be increased by simply giving the precursors?
d. When glucose is plentiful, why is glycolysis increased?
e. What is the average rate of glucose turnover in the basal state?

The circulating pool of glucose is only sufficient to maintain the brain's oxidative requirement for a few hours. Hepatic output of glucose in fasting subjects is thus an essential function. About 75 % of this glucose production comes from glycogenolysis and 25 % are derived from gluconeogenesis. Lactate makes up about 60 % of the precursors for gluconeogenesis. The source of lactate is mainly from glycolysis in muscle, erythrocytes and white cells. The other major substrate is amino acids (25 %, especially alanine). Glycerol and pyruvate can also be used in glyconeogenesis. The renal cells can also carry out gluconeogenesis.

The enzymes for gluconeogenesis are upregulated by hormones e.g. glucagon. Hepatic autoregulatory response to decreasing glucose concentration also increases the gluconeogenetic enzyme activities. The supply of precursors must be accompanied by induced enzyme action.

When the glucose is present in abundance, glycolysis is enhanced. More acetyl-CoA is formed from pyruvate and more citrate is generated in the mitochondria. Citrate diffuses back into the cytosol to stimulate the first step in fatty acid synthesis.

In the basal state, glucose turnover is about 2 mg/kg/min. Most of the use of glucose (70 %) in this state is independent of insulin. The uptake is via the constitutively-express GLUT-1 transporter.

3 Protein Metabolism

a. What is the basic daily requirement for protein in the diet?
b. Can all amino acids be used in gluconeogenesis?
c. What is the percentage requirement for essential amino acids?
d. What is meant by nitrogen balance?
e. When alanine is converted to glucose, is there a net synthesis of glucose from protein?

Daily, the protein pool turns over at a rate of 3–5 g/kg body weight. The synthesis and degenerative reactions contribute about 20 % of BMR. About 50 g of protein is irreversibly degraded daily. Thus the daily dietary need is around 0.8 G/kg. About half the spectrum of 20 amino acids are essential amino acids, which cannot be synthesized endogenously. In childhood and during infancy, 40 % of the protein intake should make up of essential amino acids.

All biochemical degeneration of amino acids converge into three metabolic pathways: gluconeogenesis, ketogenesis and ureagenesis. All amino acids, except leucine and lysine can provide carbon atoms for the synthesis of glucose. In the interdigestive period, the only source of leucine e.g. is from endogenous protein breakdown.

All amino acids are oxidizable to CO_2 and H_2O after deamination. In the steady state, the urinary nitrogen excretion (urea plus ammonia) and small nitrogen losses in feces and the skin are equal to the nitrogen produced by metabolism of endogenous and exogenous protein. During disease or tissue injury, negative nitrogen balance occurs. In a growing child, protein nitrogen intake exceeds the urinary nitrogen excretion.

In the glucose-alanine cycle, alanine functions as a carrier of amino groups from the muscle to the liver. Alanine is formed from muscle pyruvate by transamination.

In the liver, alanine is extracted and deaminated for ureagenesis. The pyruvate released then enters gluconeogenesis. Thus the carbon atoms are merely being shuttled from one glucose molecule to another via pyruvate and alanine.

4 Fat Metabolism

a. What are essential dietary fatty acids?
b. What constitutes the direct source of lipid oxidation?
c. How does the proportion of the lipoprotein components change with changes in density?
d. What are chylomicron remnants?
e. In the interdigestive period, the major source of plasma triglycerides are very-low density lipoproteins (VLDL).

Both exogenous and endogenous triglycerides consist of long-chain saturated, mainly C-16 palmitic and C-18 stearic and monosaturated, C-18:1 oleic fatty acids esterified to glycerol. Around 3–5 % of fatty acids are polyunsaturated and the body cannot synthesize these lipids (linoleic, linolenic and arachidonic).

Plasma free fatty acid (FFA) derived from adipose tissue stores or dietary fat is the direct source of 50 % of total lipid oxidation. The remainder comprises of oxidation of intracellular lipids.

The complex lipoprotein particles have varying densities from <0.94 to about 1.21. As the particle density increases (from chylomicrons to HDL), the triglyceride protein decreases. The percentage protein and phospholipid components increases from less than 5 and 10 % to 50 and 30 % respectively.

Chylomicrons have a half-life of 5 min. The endothelial cell surface of adipose tissue, muscle and heart have the enzyme lipoprotein lipase. Chylomicrons are depleted of triglycerides by the enzyme and become chylomicron remnants. These remnants are then enriched with cholesterol esters by HDL particles by the enzyme LCAT. The remnant particles are taken up by the hepatocytes and metabolize further, releasing free cholesterol. Thus the net product of chylomicron metabolism is to transport most of the dietary triglycerides to adipose tissue and the dietary cholesterol to the liver.

VLDL particles are synthesized and secreted predominantly by the liver. Initially, VLDL metabolism resembles that of chylomicrons. After lipolysis and HDL interaction, intermediate-density lipoproteins (IDL) are formed. IDL are TG-depleted and cholesterol-enriched.

5 Lipid Metabolism

a. What is the role of lecithin-cholesterol acyl-transferase (LCAT)?
b. What is the function of cholesterol-ester transfer protein (CETP)?

c. What is the density of HDL?
d. What factors affect the proportion of LDL/HDL in blood?
e. What is the major cholesterol-containing fraction in plasma?

High density lipoproteins are essential in the transfer of lipid constituents between various lipoprotein particles and also effectively between organs. HDL are synthesized in the liver and intestine. They have a half-life of about 6 days.

HDL participates in the flow of excess plasma triglycerides back to the liver. HDL also facilitates the flow of cholesterol to peripheral tissues and the reverse cholesterol transport back to the liver. The latter is significantly important in the physiological regulation of normal healthy lipoprotein plasma levels.

HDL has varying densities. The smaller and denser subparticle HDL_3 accepts free cholesterol from peripheral cells and also from IDL particles and chylomicron remnants. The cholesterol is then esterified by the plasma enzyme LCAT, activated by HDL. The cholesterol ester is then exchanged for triglycerides in other particles by CEPT, as a result the HDL becomes less dense and larger – the HDL_2 subfraction. The exchanged cholesterol is either transported to the peripheral cells as LDL or taken up by hepatocytes as part of remnant. IDL and LDL particles (reverse cholesterol transport).

LDL levels are higher in males. HDL levels are also lower in males than in females. Besides gender, smoking, obesity, lack of physical exercise and a high polyunsaturated fat diet also predispose to a higher LDL/HDL ratio (high considered as >4.0).

LDL is major cholesterol-laden particles in plasma. 50 % of LDL consist of cholesterol, most of it as cholesterol ester. HDL fraction has about 20 % cholesterol.

6 Respiratory Quotient (RQ)

Respond and elaborate on the RQ during metabolism of a mixed diet.

a. Oxidation of fats produces less CO_2 than oxidation of carbohydrate with the same amount of O_2.
b. The RQ of protein is an average of the RQs of the amino acids.
c. If the O_2 consumption is 200 ml/min, how much energy is produced / min?
d. The RQ of brain is close to 1.0.
e. If the percentage of total calories contributed by fats is 55 % at normal O_2 consumption, what is the rate of fat oxidation?

A lesser amount of CO_2 is released for the same amount of O_2 used in oxidizing fats than carbohydrate. The RQ of carbohydrate is 1 and that of fat is 0.7. Fat metabolism thus places a lesser demand on the person's ventilation. The RQ for protein represent that of all the RQs of the component amino acids.

Normally, protein is a minor source of energy. The little contribution of protein oxidation to the overall RQ can be estimated by knowing the urinary nitrogen excretion.

The kilocalories produced per liter of O_2 is quite similar for all classes of food. This is about 5 Kcal/L of O_2. At the resting O_2 consumption of 200 ml/min, this works out at $0.2 \times 5 = 1$ Kcal/min.

RQ data of individual organs reflect the metabolic processes in the tissues. The cerebral nervous tissue has an RQ of 0.97–0.99. The CNS is an obligate glucose consumer. In contrast, the muscles oxidize predominantly fatty acids at rest. During gastric acid secretion, the stomach has a negative respiratory exchange ratio (R). This is because the gastric tissue takes up more CO_2 from the arterial blood than it puts into the venous side. At normal O_2 consumption of 200 ml/min, the kilocalorie equivalent is 1 Kcal/min. For fat 9 Kcal is produced per gram. Thus, 9 Kcal/g. X g/min is equal to 0.55×1.0 Kcal/min. X g of fat oxidized per minute will be $0.55 \div 9 = 0.061$ g/min.

7 Energy Production and Stores

Elaborate on the following descriptions of energy stores in the body.

a. One mole of ATP contains about 24 Kcal of potential energy.
b. The production and transfer of energy have an efficiency of around 65 %.
c. At a given time, the total ATP content in vivo can supply energy for 1 min.
d. The caloric density of fat is 9 Kcal/g.
e. Next to fat, protein accounts for most of the total energy stores.

The basic chemical form of energy in living cells is the molecule adenosine triphosphate (ATP). The two high-energy terminal P-O bonds each has about 12 Kcal of potential energy under physiologic conditions. The P-O bonds are generated by oxidative reactions. The P-O bonds are also in continuous flux and the energy in them is utilized in mechanical work or transferred to other energy bonds in metabolic pathways.

Because the overall efficiency in the usage of the chemical energy is about 65 %, around 18 Kcal of oxidative substrate is needed to produce each P-O terminal bond of ATP (12 Kcal/bond).

The daily turnover of ATP is about 2,300 Kcal. This is equivalent to more than 100 mol of ATP or an amount of ATP in kilogram that is about equal to the body weight!

Fat constitutes 10–30 % of body weight. Fat is stored in the form of triglycerides in adipose tissue and makes up 75 % of the total caloric stores. Fat also contain little extra weight as intracellular water. Metabolically-active cytosol is less than 10 % of the adipose tissue weight.

Protein accounts for about 25 % of the potential energy resources. The component amino acid can, via gluconeogenesis, be converted to glucose. Protein however have essential structural and functional roles. Therefore catabolism of protein to provide for energy needs is a major event only in severe energy lack.

8 Energy Reserve Regulation

a. Is the number of adipose cells fixed in the adults?
b. Is total body fat genetically-determined?
c. What is the role of leptin?
d. What is the function of brown adipose fat (BAT)?
e. Do the adipocytes produce other metabolism-related factors besides leptin?

Human fat cell numbers are not fixed for life. Under humoral influences, the pre-adipocytes in mesenchymal tissue can differentiate into mature adipocytes.

About 25 % of total fat variation and 35 % of differences in subcutaneous truncal and abdominal fat seem to be genetically determined. This data is derived from studies on adopted children, monozygotic twins and the discovery in rodents of genes that produce obesity.

Leptin is secreted by adipose cells. There is a negative feedback interaction between adipocytes and hypothalamic cells. Leptin increases the cathecolaminergic discharge from the hypothalamus. The plasma leptin level and the quantity of messenger RNA for leptin correlate well with the BMI and adiposity in humans.

In rodents, an uncoupling protein in brown adipose tissue dissociate ATP synthesis from oxygen consumption. This results in thermogenesis and the use of excess calories. BAT is present in human newborns and serves to help the baby adapt to the lower ex utero temperature. In adult human though, the amount of BAT is very little. The contribution to energy regulation is not significant.

Besides leptin, adipose cells also secrete other peptides that affect metabolism. Among them is resistin that reduces the insulin-responsiveness of target cells. A type of nuclear receptor, peroxisome proliferator activator receptor (PPAR) is also found in adipose tissue. This PPAR stimulates differentiation of pre-adipocytes as well as synthesis, binding and storage of FFAs.

9 Whole Body Metabolism

Elaborate on the following aspects of metabolic physiology in the body.

a. 1 kcal is equal to 4.2 kJ.
b. Compared to carbohydrate, the amount of O_2 used to metabolize per gram of fat is larger.
c. In the adult human, the resting metabolic rate is 1.0–1.2 Kcal/min.
d. Are the carbon atoms of the different classes of food interconvertible?
e. The BMR is linearly related to lean body mass.

Energy input comes in the form of food types, carbohydrates, fats, proteins. The complete combustion of each chemical food category yields different amounts of energy. 1 kcal of energy is equal to 4184 J. The energy yield per liter of oxygen consumed is quite similar for each food class as the ratio of carbon to hydrogen atoms is alike in each food type. This is about 5 Kcal/L O_2. Per gram of foodstuff, fats provide 9.4 Kcal while carbohydrate gives 4.2 kcal.

The carbon skeletons of carbohydrate and proteins are convertible to fats. The carbon atoms of protein can be used in carbohydrate metabolic pathways. Fat carbons are hardly transformed into carbohydrates.

In the adult human, the basal metabolic rate expends 20–25 kcal per kg body weight. This is approximately 1.0–1.2 Kcal/min at an O_2 usage of 200–250 ml/min.

About 40 % of the BMR is due to central nervous system activity and 20–30 % by the skeletal muscle mass. Inter-individual variations in BMR are mainly accounted for by gender, age, fat mass and lean body mass. BMR is linearly associated with body surface area and fat-free mass.

The absolute, minimal energy in BMR is needed for a variety of cellular functions. These include (i) maintaining transmembrane ionic gradients, (ii) signal conduction in the nervous system, (iii) anabolic and catabolic biochemical reactions, (iv) mechanical work in muscle, CVS and respiratory functions, (v) heat production.

10 Control of Eating

a. Does insulin act on the hypothalamus?
b. Does leptin have inhibitory or stimulatory action on the hypothalamus?
c. Is the stomach itself involved in eating behavior?
d. Is human obesity regularly due to low plasma leptin?
e. How is body mass index calculated and used?

Insulin acts like leptin and reinforces leptin action. Plasma insulin levels are elevated when the adipose tissue is increased.

Leptin has been shown to act on at least two types of hypothalamic neurons in the arcuate nucleus. Leptin inhibits one set of neurons and stimulates the other neuron types. The neurons inhibited synthesize neuropeptide Y (NPY) which has orexigenic activity. NPY also reduce energy expenditure. Simultaneously, leptin also suppresses the production of agouti-related peptide (AGRP). AGRP is an endogenous antagonist of the anorexigenic peptide, α-MSH. The other group of neurons are stimulated by leptin. These neurons synthesize POMC/α-MSH and cocaine-amphetamine-regulated transport (CART). Both α-MSH and hypothalamic CART inhibit food intake. These second order neuropeptides then interact with ‘hunger’ and ‘satiety’ neurons in the hypothalamus. This enables the hypothalamus to coordinate feeding behavior and modulate sympathetic outflow involved in hormonal and metabolic functions. Surgical reduction of gastric volume or bypass surgery can reduce body weight by one-third over about 1–2 year period. The stretch receptor in the stomach can play a role. Recently, a new gastric hormone, ghrelin with orexigenic property has been identified. Plasma ghrelin rises during the preprandial period and drops to a minimum → around an hour after eating.

Leptin deficiency is rarely seen in human obesity. The hypothalamus appears not to receive the leptin signal indicating large energy stores. Leptin receptor resistance is a strong possibility in obesity.

BMI is the body weight in kg divided by the square of the height in meters. In Western society, a BMI > 25 is defined as overweight. As obese person has a BMI of > 27. Obesity may be associated with an increased set point for energy reserves.

11 Fasting

a. What happens to the lipoprotein lipase activity in adipose tissue?
b. What happen to the respiratory quotient?
c. What change occurs in the metabolic needs of the central nervous system?
d. What two organs provide glucose by gluconeogenesis?
e. What is the secretary rate of pancreatic insulin and glucagon?

In the fasting state, energy supply is totally from endogenous substrate. A state of catabolism is operative. Hepatic glycogenolysis can supply glucose for about 12–15 h of fasting. Gluconeogenesis also contributes to continuing glucose supply. The kidney can also carry out gluconeogenesis. The renal production contributes about 15 % of glucose after an overnight fast. The glucose precursors for gluconeogenesis comes from protein catabolism as well as lipolysis of triglycerides. Gluconeogenic amino acids and glycerol respectively are converted to glucose.

The muscle lipoprotein lipase increases to enhance the uptake of triglycerides for oxidation. In the adipose tissues, the lipoprotein lipase decreases to reduce the cellular uptake of triglyceride for storage.

The glucose oxidation in muscle and liver is spared as greater amount of FFA are available for oxidation. The metabolic shift during fasting away from glucose toward fatty acid oxidation results in a lower respiratory quotient.

These metabolic changes are effected and mediated by hormonal actions. In particular, the anabolic hormone insulin is decreased and the catabolic glucagon is elevated. After several days of fasting, the BMR decreases up to 20 %. The CNS is no longer an obligate consumer of plasma glucose. Two-thirds of its need can be met by the ketoacids which are products by fatty acid oxidation. With less demand for glucose, gluconeogenesis also slows and proteolysis declines. The contribution of renal gluconeogenesis to glucose synthesis can be up to 25 %.

12 Exercise

a. Is the increase in glucose uptake by muscle insulin-dependent?
b. What needs to be replenish to settle the oxygen debt?
c. What is the immediate energy form for muscle activity?
d. How does carbohydrate loading improve endurance?
e. What proportion of energy needs is met by fatty acids during prolonged exercise?

The major increase in glucose entry into muscle occurs by an insulin-independent way. During exercise, intracellular glucose and ATP drop and AMP rises. AMP could stimulate glucose transport by an AMP-activated protein kinase (AMPK). Hepatic glucose production is increased to maintain plasma glucose.

Within several minutes of strenuous exercise, an oxygen debt of about 12 L can accumulate. About 8 L of O_2 is needed to convert the lactic acid back to glucose in the liver or to oxidize lactic acid to CO_2. Another 2 L of O_2 is necessary to replenish normal muscle ATP and creatine phosphate stores. O_2 in body fluids and that bound to hemoglobin and myoglobin have to be restored.

A high carbohydrate diet for several days before a sustained exercise (marathon) can increase the endurance limit. This is due to the larger liver and muscle glycogen stores. In intensive physical exercise, the muscle glycogen can only supply glucose of just a few minutes.

When the exercise is prolonged, the main energy substrate becomes fatty acids (about 2/3 of the energy needs). The increased AMP leads to a decreased inhibition of carnitine palmitoyl transferase. This facilitate the entry of the fatty acids into the mitochondria for oxidation.

The profile of the substrates in plasma is similar to that during fasting except for elevated pyruvate and lactate concentrations. The latter results from the higher glycolytic activity.

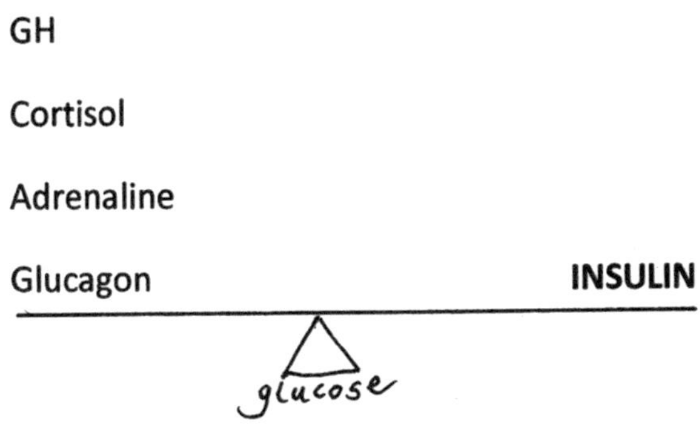

Pancreatic insulin is unique as the only hypo-glycemic hormone that lowers post-prandial blood glucose. During energy demand as in fasting, a few counter-regulatory hormones (adrenaline excluded) have hyperglycemic effect that helps to maintain blood glucose. Glucose output into blood is either via glycogenolysis or gluconeogenesis. The anabolic insulin inhibits gluconeogenesis and promotes glycogenesis.

TEACH Macarons. A special edition of Macarons made by my colleague Dr Jaclynn Ng. These macarons were given out to all candidates at the 1st Physiology Teacher's Prize event in January 2016. Jaclyn did her PhD in United Kingdom, hence the UnionJac brand.

The incentive of negative energy balance

Energy storage in comparative physiology

Homeostasis, including Energy balance has been described as Wisdom of the Body

Part IV
Integrated Physiology: Neuro, Endo, Gastroenterology

The eye cannot say to the hand, 'I don't need you!'. And the head cannot say to the feet, 'I don't need you!' Paul's Epistle to the Corinthians

A Physio Song to be sung to the popular Chinese New Year tune 'Guo Tsin Nien, Guo Tsin Nien'

Sa li vate Sa li vate

When the food is on your plate,
Pavlov's Reflex, Stimulate
Salivate, Salivate
Para sym Innervate
Blood flow increase, Vasodilate

When you chew your food, Receptors activate,
Mechano-, Chemo-, participate

Salivate, Salivate,
Ductal cells, Modulate
Sodium, K plus and bicarbonate.
Salivate, salivate
Amylase Carbohydrate
Chinese New Year, Celebrate!

When you chew your food, Receptors activate,
Mechano-, Chemo-, participate

Salivate, Salivate
Swallow, Brain Stem Locate
Peristalsis, ENS Regulate.

NEURO endocrino gastroentero LOGY

Some aspects of physiology in an ancient text "the LORD God made all kinds of trees…trees that are pleasing to the eye (neuro) and good for food (gastro)… the LORD God caused the man to fall into a deep sleep (anesthesia)… a man will be united to his wife and they will become one flesh (sex, reproduction, endocrine)

Introduction: Brain-Gut Relationships, Paracrines

Common daily scenarios are useful to illustrate the multiple, interweaving roles of neurophysiology, endocrinology and the gastrointestinal system. Let us use three examples; the baby and her mother, the boy and the blueberry cheesecake, the baldy man and his favourite barbecue.

The mother nurses her baby on demand frequently. The baby can respond to the mother's voice (auditory) and smell (olfactory) and cries to indicate her desire for mother's milk. The sensory receptors at the nipple of mummy's breast are stimulated by the baby's suckling. A neuroendocrine reflex to the hypothalamus releases the neuroendocrine oxytocin from the posterior pituitary. Oxytocin satisfies the baby's hunger by contracting the myoepithelial cells for ejecting milk from the breast. Occasionally, the cry of the baby is sufficient to trigger the mother's emotional response to produce the oxytocin action. This is the brain-breast relationship. Recently, oxytocin has been studied for its apparent 'cuddle effect' in enhancing parental-child bonding.

The boy sees (photoreception) the cheesecake in the kitchen, topped with tempting blueberry. He steals a taste, perhaps a small bite. The boy salivates even before he eats the cake, a conditioned reflex. His stomach begins to secrete some gastrin during this cephalic phase. If the time was before dinner, several hours after his lunch, the orexigenic hormone, ghrelin from his stomach is also elevated.During his dinner, autonomic parasympathetic activity predominates. Hormones that stimulate digestive juices include duodenal secretin and cholecystokinin (CCK). CCK, the key hormone stimulating bile, lipase and protease-rich pancreatic juice secretion, also activates vago-vagal reflexes on the gallbladder and pancreas excocrine cells respectively.

Generally, all hormones released from the GI tract during the postprandial period are anorexigenic, acting as feedbacks on the hypothalamic satiety centers when energy is being replenished during a meal. The undigested food, stored in the rectum is evacuated by defecation, a co-operative function of voluntary, somatic control and autonomic nervous mechanisms.

The balding male reflects his testosterone-endowed blood. Spermatogenesis is guarded within a 'blood-testis barrier'. This is equivalent to the neurons protected by 'blood - brain barrier', highlighting the importance of ensuring a constant milieu for both neuron and sperm functions. Testosterone is one of several hormones that is regulated by the hypothalamo-pituitary -endocrine organ axis.

His regular weekend pastime is to have friends over for barbecue. His 'carnivorous' diet will stimulate both glucagon and insulin simultaneously as a designed, endocrine mechanism to prevent a secondary hypoglycemia from the insulin response to elevated blood amino acids. His insulin secretion is increased by parasympathetic nerve action even before food enters the stomach. Insulin is also secreted before a postprandial rise in blood glucose, by intra-luminal stimulation of GI incretins.

Good, classical music (sound pitch discrimination) is enjoyed while neuroendocrine control of gastric emptying, sequential peristalsis are harmonized with the optimally timed, concurrent secretions into the duodenum of pancreatic bicarbonate, enzymes and gall bladder bile.

Contented, he sleeps till almost noon the next day. During the inter-digestive period, his GI tract is not quiescent. 'Housekeeping' periodic peristaltic activity, termed migrating motility complexs (MMC) 'sweeps' the gut, to prepare for the next meal. The enteric neural system (ENS) of the gut, the 'little brain' in the center of the body's physiologic universe, thoughtfully regulates the secre-motor behavior of the GI passage during both fasting and feasting.

Chapter 15
Neuro-Endocrinology

The term 'neuro-endocrine' can be used in a number of ways. Some hormones are neurohormones in that they are synthesized and released by dedicated neurons. Reflexes can be neuro-endocrine in nature, involving both a hormonal and a neural component.

The phrase 'neuro-endocrine reflex' would suggest a cause and effect that begins with a neural impulse and is completed with a hormone action. This scenario is seen in parturition in the uterine stretch > oxytocin-induced contraction. There are also situations when a hormonal stimulus can activate a neural response. The action of various gastrointestinal hormones in post prandial blood on hypothalamic neurons that regulate feeding/satiety is an example.

The neural component in the neuroendocrine homeostatic loop can be either afferent or efferent fibers. The cardiac volume receptors convey sensory, afferent signals that affect secretion of pituitary vasopressin. In the volume/baroreflex response, it is the efferent renal sympathetic that stimulates release of renin. Any increase in general sympathetic activity during emotional responses to stressful conditions will trigger release of catecholamines from the adrenal medulla that has cholinergic sympathetic innervation.

To add to the physiologic variations in nerve and hormone interactions, neurohormones can determine the secretion from other endocrine cells. This is clearly the picture in the brain where the hypothalamic releasing/inhibiting neurohormones control anterior pituitary trophic hormone secretions.

Neuroendocrine physiology is a network integrating the hardwired neural pathways with the 'wireless', circulating hormones in 'vascular space' and the local paracrines in the cellular neighbourhood.

H.M. Cheng (ed.), *Physiology Question-Based Learning*,
DOI 10.1007/978-981-10-0877-1_15

1 How Are Autonomic and Neuroendocrine Reflexes Involved in Blood Volume Control?

Answer The control of blood volume involve two major hormones, renal renin and pituitary vasopressin. Renin secretion is stimulated by renal sympathetic nerve and vasopressin release is modulated by sensory afferent impulses from vascular volume sensors.

Concept Blood comprises the cellular volume; red blood cells (the hematocrit), leucocyte and platelet populations and the plasma fraction. The hematocrit, dependent on the red cell count is homeostatically regulated by the hormone erythropoietin that is secreted when the endocrine renal cells senses tissue hypoxia.

Blood volume is maintained also by the hormonic enzyme, renin which is released from the juxta-glomerular cells at the renal afferent arteriole. On major excitatory input for renin secretion is renal sympathetic nerve. Autonomic reflexes that participate in blood volume/pressure regulation are effected via changes in the general sympathetic discharge. Volume sensors like the arterial baroreceptors and volume receptors in the heart /pulmonary vessels are part of the autonomic reflex loops that activate the compensatory sympathetic activity.

Hypovolemia are sensed by the baroreceptors and renin is secreted with an increase in the renal sympathetic nerve action as part of the general increase in the autonomic sympathetic activity. Renin, once secreted leads to production of anti-natriuretic hormones like angiotensiin II and aldosterone, that conserve sodium by reducing urinary sodium excretion. The sympathetic nerve constricts the renal arterioles and this has the effect in decreasing the filtered sodium load. Vasoconstriction of the renal arterioles also helps raise the total peripheral resistance to compensate for the hypotension/hypovolemia.

Volume sensors in the heart chambers as well as the baroreceptors are also involved in the neuroendocrine reflex that influences the secretion of the hypothalamic hormone, vasopressin from the posterior pituitary. There is a level of tonic afferent impulses from the volume sensory receptors that signals the degree of vasopressin secretion. Hypovolemia results in more vasopressin release that will act in the kidneys to conserve water. Vasopressin as its name suggests is also a vasoconstrictor.

The mechano-receptors in the cardiac wall are also linked to the secretion of the cardiac hormone, a natriuretic peptide that both the atria and ventricles produce. When the ECF and blood volume becomes expanded, cardiac natriuretic hormones are increased in the blood to act in the kidneys to increase sodium excretion. Renal paracrine natriuretic factors like urodilatin are also secreted.

The autonomic reflex that are triggered by reduction in vascular volume also increases the cardiac sympathetic activity to produce compensatory tachycardia and increase myocardial contractility to improve the cardiac output. Selective sympathetic constriction of vascular smooth muscles at regional arterioles that supply the splanchnic and cutaneous circulations also helps to redirect and sustain adequate cerebral blood flow during hypovolemia.

The sympatho-adrenomedullary pathway is also activated during the autonomic reflex triggered by hypovolemia. The catecholamines, adrenaline and noradrenaline, released into the circulation augment the compensatory sympathetic neural actions.

2 How Are Muscle Functions in the Body Controlled by Nerves and Hormones?

Answer Skeletal muscles require action potential stimulation in alpha motor neurons. The smooth and cardiac muscles have extrinsic autonomic innervations and both muscle types also respond to specific hormones. Skeletal muscle contraction is not under any hormonal control.

Concept The skeletal muscle is the effector in the somatic nervous system. The final common pathway to the skeletal muscles is the alpha motor neuron that supplies the functional motor unit in the muscles. The cholinergic alpha motor neuron is always excitatory when stimulated, producing the muscle movement by excitation -contraction coupling mechanisms. Hormones are not needed and are not involved in the skeletal muscle contraction events. Hormones like adrenaline do increase skeletal blood flow during physical activity by a vasodilatory action. The steroid hormones, androgens increase muscle mass through increased protein synthesis, a long term and not an acute change.

In contrast smooth muscles are responsive to hormones. For example, the vascular smooth muscles either constrict or dilate when acted upon by circulating vasoactive agents. These agents in blood commonly produce effects in the vascular smooth muscles via stimulating vasodilator or vasoconstrictor paracrines released from the endothelial cells. The muscle fibers in the single unit or visceral smooth muscle group that are electrically connected as a functional syncytium by gap junctions have extrinsic autonomic innervations. The innervating fibers course along the smooth muscle tissues and comes into synaptic contact with multiple fibers via varicosities that contain the neurotransmitters.

Depending on the tissue, the sympathetic nerve either excites or inhibits the smooth muscles. For example, the arterioles are vasoconstricted by sympathetic adrenergic nerves but the gastrointestinal (GI) tract smooth muscle walls are relaxed by sympathetic action. However, sphincters in the GI tract have higher tone with sympathetic stimulation like the internal sphincter of the bladder and the anus. The visceral smooth muscle of the GI tract also exhibit spontaneous basal electrical (BE) 'slow wave' activity, that are independent of the autonomic nerves. Interstitial cells of Cajal are likely to be the GI pacemaker cells that produce this rhythmic electrical phenomenon. Hormones or parasympathetic stimulation that stimulate GI motility will increase the generation of action potentials from this BE.

The action of hormones on smooth muscles to produce contraction, independent of action potential generation is termed pharmaco-mechanical coupling. This hormone action increases cytosolic calcium by opening membrane calcium channels either via the receptor binding or through IP_3 signalling that triggers calcium release from the sarcoplasmic reticulum.

The cardiac muscle is auto-rhythmic and its contractility is directly increased by cardiac sympathetic nerves. Circulating catecholamines augment the positive beta adrenergic action on the heart by noradrenaline that is released from the sympathetic nerves. Recruitment of motor units or recruitment of more alpha motor neurons increases skeletal muscle contraction. However in cardiac muscles, an increase in intracellular calcium by sympathetic or hormonal excitation is the positive inotropic factor. The neuro-hormonal stimulated increase in cardiac muscle metabolism during physical activity produces metabolite vasodilators that increase the coronary blood perfusion in parallel with the cardiac work load.

3 How Are Sensory Receptors and Neuroendocrine Mechanisms Involved in Controlled Gastric Empyting?

Answer Diverse sensory receptors in the duodenum including mechano-, chemo-, and osmoreceptors monitor the gastric chyme and effect regulated gastric emptying via neural and hormonal mechanisms.

Concept The stomach stores food and partial pre-digestion takes place mainly by gastric pepsin. The introduction of chyme into the duodenum for further complete digestion of carbohydrates, proteins and lipids is a controlled event. Gastric emptying occurs at an appropriate rate so that duodenal digestion can proceed optimally without being overloaded especially with excessive acidic, fatty or hyperosmotic food.

There are a variety of duodenal receptors that detects different parameters in the gastric chyme. These parameters include osmolarity, acidity, fatty acid and proteins. The sensing of osmolarity is related to preventing hyperosmotic chyme from interfering with water reabsorption in the intestine. The scenario of 'dumping syndrome' (not 'dumPling'!) occurs where reverse entry of fluid into the gastric lumen is driven by hyperosmotic contents. Hypovolemia results. The effector pathway of the duodenal osmoreceptor reflex likely involves local enteric neurons.

Chemoreceptors that are sensitive to acidity also play a part in inhibiting gastric emptying. This response has a dual purpose of protecting the duodenum from acid erosion and ensuring an optimum pH for digestive enzymes. Duodenal acidity might provoke a short enteric neural reflex that decreases the gastric motility. In addition, the release of the hormone secretin by acid leads to an alkaline pancreatic juice secretion. The pancreatic enzymes are not effective in acidic medium and neutralization of the acidic chyme by pancreatic bicarbonate is needed. Secretin supplements the neural inhibition of gastric movement by also reducing the rate of gastric emptying .

Both fatty acids and peptides stimulate duodenal cholecystokinin (CCK) secretion. The hormone CCK then increases bile flow into the duodenum by contracting the gall bladder. CCK also stimulates secretion of a pancreatic enzyme-rich juice that includes lipase. Since the emulsification of lipids to facilitate lipase action is quite an involved process, CCK has a physio-logical inhibiting action on gastric emptying. This gastric effect by CCK could be a direct hormonal action or via possible activation of sensory afferent vagal fibers in a vago-vagal reflex.

Over-distension of the duodenum by food bolus also stimulates mechanoreceptors. Local enteric and long-loop vagal reflexes will then mediate a slowing down of further entry of gastric chyme.

4 How Is the Sympathetic-Adrenal Axis Involved in Glucose Homeostasis?

Answer Sympathetic nerve action to the pancreatic islet inhibits insulin and stimulates glucagon secretion. Adrenaline from the adrenal medulla, together with sympathetic activity stimulate lipolysis in adipose tissues and also hepatic glycogenolysis and gluconeogenesis.

Concept The functions of the autonomic sympathetic neural division is sometimes described as 'catabolic'. This relates to sympathetic 'fight or flight' situations when the demand for energy supply is a priority.

The sympathetic nerve actions to its target tissues together with the release of sympathomimetic adrenaline by cholinergic sympathetic fibers at the adrenal medulla collaborate to produce an increase in blood glucose. In other words, the overall sympathetic action is hyperglycemic.

At the islet of Langerhans in the endocrine pancreas, sympathetic increases the secretion of the hyperglycemic glucagon but inhibits the body's only hypoglycemic hormone, insulin. Glucagon is catabolic and insulin is anabolic (insulin sounds like 'insurance'!). So blood glucagon level is higher in 'fight or flight' scenarios.

In adipocytes, adrenergic receptors are acted upon by both noradrenaline released from sympathetic nerve or by circulating adrenaline. The activity of hormone sensitive lipase (HSL) is activated and lipolysis of stored triglycerides leads to a rise in blood fatty acids and glycerol. The fatty acids can be used as energy substrate and the glycerol can be converted to glucose by hepatic gluconeogenesis. The adipocyte HSL is inhibited by insulin during the post prandial period when fresh energy stores are absorbed, including fatty acids and triglycerides.

Both adrenaline and sympathetic act at the hepatocytes to increase gluconeogenesis. Hepatic and skeletal glycogenolysis are also increased by adrenaline. Sympathetic nerve input to the liver causes more hepatic output of glucose from glycogenolysis.

Hypoglycemia can be viewed as a stressor. Glucose sensors in the central nervous system respond to hypoglycemia by activating a general increase in sympathetic discharge. The sympathetic nerve acts on the liver and adipose tissues. The concurrent release of adrenaline stimulated by sympathetic nerve. Hypoglycemia also stimulates pancreatic glucagon secretion.

The catabolic, net hyperglycemic effects of sympatho-medullary system ensures adequate glucose for cerebral neurons. Indirectly, the use of fatty acids and gluconeogenic glycerol from lipolysis as the energy options 'spares' the body's protein. So we could say that the autonomic system has the homeostatic role in being 'sympathetic' to proteins in the tissues of our body.

5 How Are Neuroendocrine Mechanisms Involved in the Energy Homeostasis During Exercise?

Answer The mobilizing of metabolic fuels during exercise by neuro-endocrine mechanisms is similar to that in the fasting, postabsorptive inter-digestive period. The differentiating event is the increased uptake of glucose by the active muscles while there is an adaptive decreased peripheral utilization of glucose in the fasting individual.

Concept During fasting, the sympathetic nerve activity is increased by hypoglycemia, probably detected by central glucose sensors. In exercise however, plasma glucose is little altered during mild, moderate activity and only in prolonged, intense exercise does the blood glucose drop significantly. The sympathetic activity that produces tachycardia and increased cardiac output during exercise is centrally activated, independent of blood glucose concentration.

The metabolic pattern of an exercising person includes increased hepatic glucose output both by glycogenolysis and gluconeogenesis. Increased adipose tissue lipolysis of stored triglycerides supplies fatty acids for the active muscles and glycerol is converted to glucose in the liver. The hormonal effectors of this energy shift are similar to that in the fasting person. There is reduced insulin and increased glucagon secretion during exercise. The decreased insulin/glucagon ratio reflects a catabolic status and the need for energy mobilization.

The lowered insulin/glucagon ratio results from direct sympathetic actions in the pancreatic endocrine islet cells and to elevated circulating adrenaline, secreted in response to activity in the sympatho-adrenal axis. In more intense exercise, hypoglycemia also directly increases glucagon and decreases insulin secretion. Growth hormone is also increased by hypoglycemia during fasting. Is there a separate stimulus for secretion of the hyperglycemic growth hormone during moderate exercise?

Cortisol has an essential permissive effect for gluconeogenesis and lipolysis to occur. The 'anti-insulin' effect of cortisol helps maintain blood glucose level by decreasing peripheral uptake and utilization of glucose. It is suggested that the amino acids that are released by a combined effect of increased cortisol and decreased insulin on protein catabolism, especially during prolonged exercise, is a preparatory adjustment for tissue repair if injury should happen. Gluconeogenic amino acids also contribute to the energy provision.

Interestingly, there is obvious increased glucose uptake by the metabolically active muscles during exercise even though the insulin-activated facilitated glucose transport is reduced. There is certainly an insulin-independent pathway during exercise for the increased cellular extraction of blood glucose. The recommendation of regular exercise in diabetics for glucose control has this cell physiological basis.

6 How Would the Ventilation/Perfusion Matching Be Tied in with Associated Brainstem Activities that Effect Neuro-Endocrine Homeostatic Mechanisms

Answer Pulmonary ventilation is controlled by respiratory neurons in the brain stem. Pulmonary perfusion is the right ventricular cardiac output (CO). Cardiovascular regulatory neurons in the brainstem medulla activate neuro-endocrine pathways to maintain blood volume and also blood pressure, of which CO is a determinant.

Concept Thinking integratively in Physiology is a useful exercise to visualize and appreciate the big picture of homeostasis in our body. Some integration are more obvious e.g. cardio-respiratory system in delivering oxygen to the cells.

This question above requires a broad synthesis of several physiological systems but is useful to trigger and help students to recall essential functional concepts in different organ physiology.

The ventilation /perfusion (V/Q) balance described in respiratory functions involved matched airflow to blood flow, considered regionally and for the lungs as a whole. The brainstem is the location of the respiratory pacemaker neurons (pre-Botzinger complex) and the basal neuronal activity is influenced by sensory inputs from various sectors e.g. arterial chemoreceptors.

The pulmonary blood flow is the ventricular output from the right ventricle. Cardiac output is influenced by blood volume. CO is then a determinant of arterial blood pressure. When we review blood volume/pressure regulation, cardiovascular reflexes involving the brainstem come into view.

The efferent sympathetic outflow from baro/volume receptor reflexes integrated by the cardiovascular brainstem neurons is involved in both the rapid response and long term compensation of blood pressure.

Activation of the sympatho-adrenal axis increases circulating catecholamines that enhances the sympathetic actions on the heart, arterioles and veins. The specific sympathetic nerve to the kidneys effects sodium reabsorption and maintains blood volume, directly by reducing filtered sodium load and indirectly through the family of hormones in the renin-angiotensin system.

Pulmonary blood flow is 5 L/min in a male, 70 kg adult. The alveolar ventilation at all the pulmonary gas exchange areas total around 4 L/min. The overall V/Q ratio for the lungs is 0.8. You could then make a big picture statement by saying that normal brainstem function contributes to the normal V/Q since both ventilation and pulmonary perfusion involve respiratory and cardiovascular centers in the brainstem respectively.

Students can also recall that the pulmonary vessels generate angiotensin II for blood volume/pressure control.

Recently, there are interesting research findings that the respiratory chemoreceptor reflexes can determine a 'crossover' effect on the brainstem sympathetic cardiovascular responses that affect blood pressure.

7 What Are the Neuro-Cytokine Mechanisms that Are Involved During Fever?

Answer Circulating endogenous pyrogens (EP) include some inter-leukins that are produced during infections. Induced hepatic production of EP activate afferent vagal fibers that also cause fever by resetting the hypothalamic thermocontrol set point.

Concept The body's core temperature is tightly regulated over a narrow normal range. Both central and peripheral thermoreceptors are sensors that detect and trigger appropriate heat loss or heat gain compensatory mechanisms during hyperthermia or hypothermia respectively. The integrative thermo HQ neurons are in the hypothalamus.

It should be noted that in the hyperthermia of exercise, with continuous physical activity, the core temperature stabilizes at an elevated value . This equilibrium reflects the balance between increased heat production by the active muscles and a higher heat loss reflexes, orchestrated by the hypothalamus.

In the fever scenario, there is no general abnormality in the temperature homeostatic mechanisms. The elevated core temperature is due to a resetting of the hypothalamic thermostat to a higher 'normal' value.

Most commonly, fever is associated with infections. A collective group of chemicals, released during infection and generates the fever has been termed 'endogenous pyrogens'. Three established EP are interleukin 1, interleukin 6 and tumor necrosis factor alpha.

These EP are increased in circulation, released from the interaction of the infectious microorganisms with the immune cell, macrophage. Within the liver, the secretion of EP by macrophages also activate the sensory fibers of vagal afferents. The blood-borne EP and the activated hepatic vagal both change the thermostatic control set point in the hypothalamus to a value above 37 °C. One converging final mediator of the new 'normal' fever temperature is hypothalamic prostaglandins. The anti-pyretic action of aspirin is effective as such.

The 'chills' and heat-gain shivering responses during fever is explained by this resetting of the hypothalamic thermostat. Since the EPs also have action that contributes to resistance against infection there is a conceptual idea that, fever a constant companion of infections may serve a protective, physiologic role.

Recent findings have included additional factors that also play a part in thermoregulation. Chemical messengers, called 'endogenous cryogens' are released by the hypothalamus and other tissues to prevent excessive elevation of core body temperature. These 'cryogens' could help to restore the reset thermostat back to normal when the fever inducing agent is removed. Hormones like vasopressin, the glucocorticoids and interleukin 10 are among such 'cold' chemo-candidates.

8 How Is the Secretion from the Pineal Gland Under a Neuro-Endocrine Reflex Control?

Answer The pineal gland secretes melatonin, which is entrained to the light–dark cycle in the environment via a reflex involving the retino-hypothalamic nerve fibers.

Concept The pineal gland arises from the roof of the third ventricle. The pineal gland contains the enzymes for converting serotonin to melatonin. In humans, the synthesis and secretion of melatonin are at low levels during the daylight hours but increased during the dark phase of the day.

This interesting diurnal hormonal secretory rhythm of melatonin is due to similar profile of action potentials in the post-ganglionic sympathetic nerves that innervate the pineal gland. The noradrenaline neurotransmitter acts on beta adrenergic receptors in the pineal endocrine cells. Intracellular cyclic AMP is increased which then heightens the activity of the enzymes that convert serotonin to melatonin in the pineal.

In turn, the pattern of diurnal sympathetic discharge to the pineal gland is synchronized to the light -dark cycle via the afferent sensory impulses in the retino-hypothalamic fibers that terminate at the suprachiasmatic nuclei.

From the hypothalamus, descending fibers converge on the intermediolateral gray column of the thoracic spinal cord and synapse with pre-ganglionic sympathetic neurons.

These neurons in turn synapse with post ganglionic sympathetic neurons in the superior cervical ganglion.

Pineal functions including melatonin action might have a role in the timing and initiation of puberty because pineal tumors are sometimes associated with sexual precocity.

Melatonin should not be confused with melanin which are found as granules in cutaneous melanocytes in mammals. The melanocytes contain receptors for melanocyte-stimulating hormone (MSH). However, MSH from the anterior pituitary do not circulate in humans but the precursor for MSH, pituitary adrenocorticotrophin (ACTH) binds to the receptors. Thus hyperpigmentation in primary adrenal insufficiency is due to the elevated levels of ACTH.

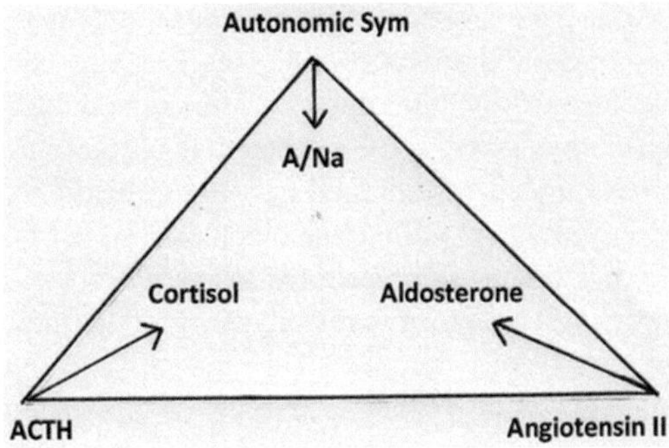

The adrenal glands and the sympathetic autonomic nerve. The catecholamines, adrenaline and noradrenaline are secreted in response to cholinergic sympathetic nerve activity to the adrenal medulla. The enzyme for conversion of noradrenaline to adrenaline are upregulated by cortisol from the adrenal cortex. The hypothalamus regulates cortisol secretion via its effect on anterior pituitary ACTH. The hypothalamus is also a central participating center for autonomic neural activity. The other major corticosteroid from the adrenal cortex is aldosterone, which acts together with cortisol in maintaining fluid balance, blood volume and arterial pressure.

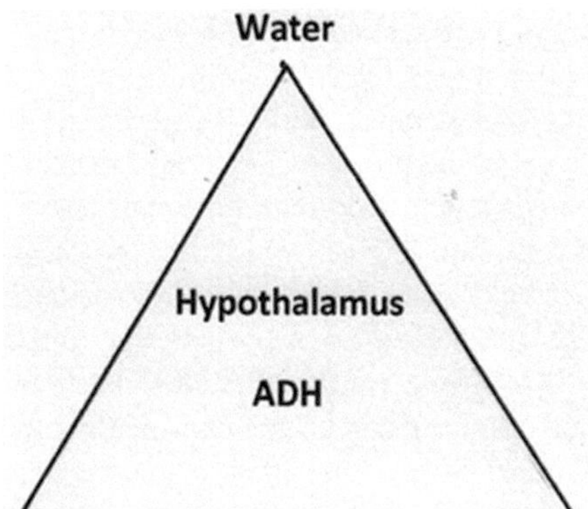

The physiology of water balance, osmoregulation and control of ECF sodium concentration are functionally synonymous. Sodium concentration is the primary determinant of osmolarity. The osmoreceptors are in reality sodium concentration sensors in the hypothalamus. Clinically, changes in osmolarity are almost always due to changes in water balance rather than a result of gain/loss in total body sodium

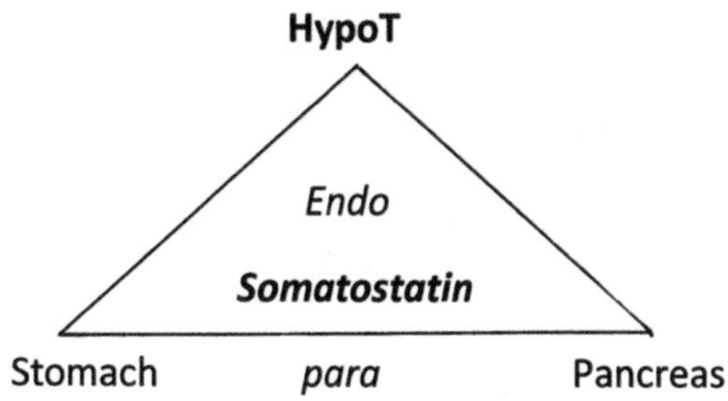

The somatostatin peptide is both an endocrine and a pararine hormone. The hypothalamic somatostatin inhibits growth hormone release from the anterior pituitary. Gastric somatostatin is increased by low pH in the empty stomach and this local somatostatin suppresses hydrogen ion secretion via gastrin inhibition. Somatostatin is a paracrine regulator for insulin and glucagon secretion in the pancreatic islet of Langerhans.

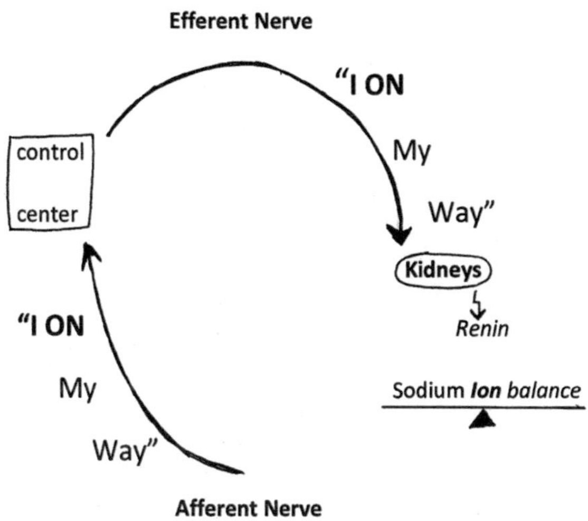

Physiology is often a moving, ionic event! Nerve and Ions, in particular sodium cations. Sodium fluxes are the ionic generators of action potentials along both afferent sensory and efferent nerves. The renal sympathetic nerve stimulates renin secretion which activates a series of hormones, including aldosterone that preserves sodium ion balance. This neuro-ionic control scenarios is also seen in the insulin control of potassium in post prandial blood. Parasympathetic, vagal fibers increase insulin secretion from pancreatic endocrine cells during a meal. Insulin action is hypokalemic by increasing cellular uptake of potassium.

Chapter 16
Neuro-Gastroenterology

The gastrointestinal (GI) system extends from oral end to anal end and is in contact with and part of the external environment. The 'brain of the gut' or the enteric nervous system (ENS) is stretched out as an integrating neural matrix within the wall of the GI tract. The total neurons that populate this ENS is estimated to be comparable to that in the CNS.

The GI tract is also functionally the 'largest' endocrine organ and it secretes a variety of hormones that have not merely pro-digestive roles. The specific GI endocrine cells are clustered and localized along different segments of the gut.

A major neural activity in neuro-GI functions is the autonomic parasympathetic nerves. These are largely branches of the wandering vagus nerve that supply the splanchnic vasculature and increases secreto-motor activities during feeding. Remember 'feasting at Las Vagus'. Parasympathetics increase insulin, gastrin and bile secretions for carbohydrate, protein and lipid GI handling respectively.

The ENS of the GI system are functionally linked to the CNS via the autonomic nervous system. Persons who suffer from ulcers due to high-demand jobs and patients with problems of stress-induced diarrhoea demonstrate this brain-gut interconnectivity.

A denervated gut might still peristalsis along with its ENS intact but the absence of co-ordinated, motility and secretory patterns (including the vago-vagal reflexes from different GI segments) will lead to inadequate digestion of a meal.

There are also hormonal and afferent neural signals from the GI during eating, when anabolism predominates, that provide satiety signals in the overall homeostasis of energy balance.

H.M. Cheng (ed.), *Physiology Question-Based Learning*,
DOI 10.1007/978-981-10-0877-1_16

1 How Does Defecation Involve Both Smooth and Skeletal Muscle Function?

Answer The anal sphincter has an internal and external component. The external sphincter is a skeletal muscle, innervated by the pudendal nerve. The smooth muscles of the internal sphincter are relaxed, while the rectal and the sigmoid colon smooth muscles are contracted by parasympathetic nerve during defecation.

Concept The rectum is normally empty. Distention of the rectum by feces that enter by mass movements from the colon initiate the defecation reflex. Mass movements are slow moving but powerful contractile waves that move over large areas of the colon three to four times each day. Distention of the stomach by food can increase the colonic motility and mass movement (gastro-colic reflex). This reflex that produces a desire to defecate after a meal is enhanced by the hormone gastrin.

The gastro-ileal reflex complements the gastrocolic response by moving any remaining small intestinal contents into the large intestine.

The internal anal sphincter is supplied by both an excitatory sympathetic and an inhibitory, parasympathetic nerves. During the inter-defecation phase, the internal sphincter is tonically contracted. The external skeletal sphincter tone is also maintained by impulses in the pudendal ('poo' dendal nerve!). This achieves the normal fecal continence.

Filling of the rectum causes relaxation of the internal anal sphincter via enteric intrinsic nerves that release VIP or nitric oxide. This rectoanal reflex is opposed by concurrent increase in the external anal sphincter tone.

Before increased rectal pressure relaxes the external sphincter, voluntary defecation can be aided by a forced expiratory effort in 'straining at stools'. This represents a Valsalva's maneuver to increase the intra-abdominal pressure by voluntary contractions of the diaphragm and abdominal muscles against a closed glottis.

The angle between the anus and the rectum is normally about 90°. During defecation, the muscles of the pelvic floor relax and this reduces the ano-rectal angle to 15°. For the unsocial behavior of voluntary expulsion of flatus, the contractile activities are similar but the pelvic floor muscle does not relax and there is no change in the rectoanal angle. This allows the gaseous flatus to be forced past the acute angle of the anorectum without the undesired simultaneous exit of feces!

Delayed defecation leads to constipation. Feces is hard and dry due to a greater than normal absorption of water as a result of long transit colonic time. Delayed expulsion of feces may be due to (i) poor defecation habits (ii) low fiber diet, aging or emotions (iii) obstruction in the large intestines (iv) lesion in the defecation reflex neural pathways.

2 How Are Gastrointestinal Reflexes Co-ordinated?

Answer Gastrointestinal (GI) reflexes are of two major types; local, short reflex arcs mediated via the enteric nervous cells and long reflexes that involve the central nervous system. These homeostatic GI reflexes do not regulate the constancy of the extracellular fluid (internal environment of all cells) but are governed by the GI luminal status during feeding and the inter-digestive period.

Concept Gastrointestinal reflexes are initiated by the volume and composition of the food or chyme. Distension of the GI wall are sensed by stretch mechano-receptors. The various nutrients in digested food are sensed by chemo-receptors that are responsive to fatty acids, peptides, monosacchrides, as well as pH. There are also duodenal osmoreceptors that e.g. regulate gastric emptying by neuro-hormonal reflex mechanisms to ensure optimal iso-tonicity of the chyme to allow for physiologic vectorial water reabsorption into the blood circulation.

In general, the neurons in the myenteric plexus of the enteric nervous system (ENS) project to affect smooth muscle activity of the GI wall and the submucosal plexus effects secretory actions. The local ENS thus monitors via sensory receptors the physiological status of the lumen, both in the digestive period and in fasting. Sensory information are also conveyed to the central nervous system (CNS) and effector responses are received and integrated by ENS neurons to cause the appropriate secretomotor and vascular responses. The vago-vagal long reflex loops participate e.g.in both gastric acid, pancreatic juice and gall bladder bile secretions.

Initiation of the GI reflexes are not restricted to sensory receptors within the GI tract. The cephalic responses from the sight (photo receptors) and smell (olfactory receptors) of food provide a measure of preparatory salivation, gastric juice and pancreatic juice release.

GI reflexes are also integrated with the homeostatic mechanisms of energy balance. For example, the gastric hormone ghrelin is orexigenic and is secreted from the empty stomach to increase the hunger sensation for more energy input. Intestinal peptide YY in contrast is anorexigenic and inhibits the desire for eating. The decrease in appetite after radical gastrectomy for treatment of obesity may be contributed by the early release of anorexigenic mediators like peptide YY from the intestine (students; do not confuses this with neuropeptide Y, a hypothalamic neurotransmitter that stimulates feeding).

The presence of glucose in the intestinal lumen reflexly signals, via several GI hormones collectively called 'incretins', to increase pancreatic insulin release even before any measured rise in postprandial blood glucose. This feed-forward insulin response prepares for the impending hyperglycemia during feeding.

Another example of GI reflexes determined by luminal triggers is the activity of the periodic, migrating motor complex (MMC) that is observed in the fasting state. The MMC has a housekeeping function for intestinal health to prevent accumulation of undigested food from a previous meal. In addition, the MMC might also prevent reflux of colonic bacterial contents into the ileum.

In the presence of food, the MMC is interrupted and the peristaltic, mixing gastric and intestinal contractions take over. The GI fed motility pattern is not seen by infusing the nutrients parenterally. The changeover in GI smooth muscle activity from the fasting MMC is thus generated through GI reflexes, both short and long, triggered by the presence of luminal dietary nutrients.

3 How Do Neural Dysfunctions Account for the Symptoms in Ileus, Irritable Bowel Syndrome and Hirsprung Disease?

Answer Ileus in the traumatized intestine is due to direct reflex inhibition of smooth muscle function. Hirschsprung disease is due to a congenital absence of ganglion neurons in the enteric nervous system. Irritable bowel syndrome is believed to be due to a visceral hypersensitivity.

Concept After an abdominal surgery, intestinal activity is temporarily suppressed. Intestinal peristalsis is restored in about 8 h, then gastric peristalsis but colonic motility can resumed much later after 2 days.

The adynamic ileus can be minimized by the use of minimally invasive e.g. laparoscopic surgery. The smooth muscle inhibiton is partly due to activation of local opioid receptor binding in the intestines. Opioid antagonists have been studied in clinical trials to assess this potential to restore intestinal motility. When the peritoneum is incised, there is a reflex increase in noradrenergic fibers in the splanchnic nerve. This autonomic sympathetic activity inhibits intestinal motility.

Abnormal colonic activity in Hirschsprung disease or aganglionic megacolon presents with abdominal distension, and anorexia. There is a developmental absence of both myenteric and submucosal plexus in the segment of the distal colon. Also found is a deficiency of the interstitial cells of Cajal. The absence of peristalsis results in a poor movement of the undigested feces over the aganglionic gut portion. Children with Hirschsprung's may only defecate once in 3 weeks. Surgical resection of the aganglionic segment and anastomosis of the remaining normal colon to the rectum restores normal colonic transit time and defecation.

Mutations in the receptors for certain endothelins or glial-derived neurotrophic factor (GDNF) appear to account for the failure of embryological neural cell migration and the lack of colonic ENS plexuses. The normal colon extracts about 90 % of the fluid in the isotonic chyme that enters it from the ileum to leave behind the semisolid feces.

Functional bowel disorders including irritable bowel syndrome (IBS) belong to a broad group of gastrointestinal problems which have no clear organic cause. The patients complain of abdominal pain, diarrhea and/or constipation and bloating. There is commonly a heightened gastrocolonic reflex, following a meal. The received wisdom and knowledge currently is that there is a visceral hypersensitivity in IBS. Normal physiological distensions of the gut are perceived as a painful event. The hypersensitivity might result from a modulation of afferent fibers from the GI tract.

The frequent diarrhea encountered in ulcerative colitis seems to be due to a different causative mechanism; allergic, immune-mediated or a chronic bacterial infection. Large areas of the colon are inflamed and ulcerated and mass movement motility are more frequently stimulated throughout the day. A temporary ileostomy is sometimes performed to allow the ulcerated colon to heal.

4 How Is the Control of Micturition Involving Somatic and Autonomic Mechanisms, Similar to that of Defecation?

Answer The voiding of urine is also regulated via a skeletal external urethral sphincter and autonomic actions on the smooth muscle of the bladder and the internal urethral sphincter. Brain stem pontine and midbrain neurons also have descending inputs onto the micturition spinal reflex.

Concept The ureters fill the bladder via periodic peristaltic contractions of the smooth muscles of the ureter walls. There are no ureteral sphincters but the oblique passage of the ureters into the bladder prevents reflux of urine from the bladder. The epithelia that line the ureters and the bladder are impermeable to water and solutes and urine composition is not modified as occurs for bile in the gall bladder.

Smooth muscle plasticity and physical mechanisms of Laplace Law account for initial bladder filling without a great increase in the intra-vesical pressure. The reciprocal autonomic innervations by sympathetic and parasympathetic arms control bladder emptying.

During bladder filling, sympathetic motor activity in the hypogastric nerve dominates. The circular muscle of the bladder, the detrusor muscle is relaxed via a beta receptor action and this occurs at the same time the internal sphincter is contracted by an alpha receptor effect. The internal sphincter is actually the bladder neck or posterior urethral wall that exhibits a high tension. The external sphincter is contracted by entrained, voluntary action via somatic motor fibers in the pudendal nerve. These reflexes are guarding reflexes that serve normal urinary continence. The normal sense or urge to urinate as the bladder fills can be voluntarily suppressed by the combined Somatic/Sympathetic (SoS) nerve actions.

As the bladder fills, the fullness is sensed by stretch, mechanoreceptors in the bladder wall. Afferent impulses are relayed to the spinal cord and the brainstem. The voiding of urine occurs when descending fibers from the brainstem override the guarding spinal reflexes. There is a pontine area that has neurons that facilitate the micturition reflex. Parasympathetic activity predominates with contraction of the detrusor muscles and relaxation of the internal sphincter via cholinergic muscarinic actions. Both fibers of the afferent and the efferent limbs in the parasympathetic voiding reflex travel in the pelvic nerve.

If the afferent sensory fibers from the bladder are damaged, the bladder becomes atonic as the micturition reflex is abolished, even though the efferent fibers are intact. The bladder is distended and urine is lost as a dribble through the urethra.

In spinal shock, when the control of descending fibers are lost, spinal bladder reflexes eventually returns. Because the voluntary control is absent, an automatic bladder results. Micturition is triggered and urine voided when the intra-vesical pressure reaches a threshold level. There is no constant urine dribble. This is similar to the scenario in babies before they have learnt wee-lingly!

The smooth muscles surrounding the urethra appear to be less important in voiding of urine but might function rather in preventing or 'avoiding' the reflux of semen during ejaculation.

5 How Do Action Potentials in Nerves Promote Protein Digestion in the Intestines?

Answer During the cephalic, gastric and intestinal phases of protease-rich, pancreatic juice secretion, neural stimulation including parasympathetic vagal, vago-vagal reflexes, combined with CCK, Serotonin and Secretin actions enhance and potentiate pancreatic exocrine secretions.

Concept Quantitatively, the pancreatic secretory response is about 70 %, 20 % and 10 % during the pancreatic, cephalic and gastric phases respectively. Cholinergic vagal stimulation occurs with the sight, smell, taste and chewing of foods. Gastric distension activates vago-vagal reflexes that stimulate, as in the cephalic phase, a low flow, high digestive enzyme concentration pancreatic secretion. This response reflects a primarily pancreatic acinar cellular stimulation.

In the intestinal phase, the pancreatic ductal cells are also strongly stimulated. The ductal bicarbonate-rich secretion alkalinizes and dilutes the final secretion that drains into the duodenum. The duodenal hormone secretin binds to receptors on the basolateral pole of the epithelial ductal cells to produce this action. Entero-pancreatic reflexes also contribute to the large volume of pancreatic secretion when chyme enters the duodenum. Both CCK from duodenal endocrine and serotonin (5 HT) from intestinal enterochromaffin cells act on the acinar cells.

It has been estimated that CCK and serotonin each accounts for around 50 % of the pancreatic enzyme secretions during the intestinal phase. Interestingly, CCK appears to act predominantly via stimulation of vagal afferents near its site of release in the duodenum. The vagal efferents that release acinar digestive secretions are cholinergic fibers as well as VIPnergic and GRPnergic finers. A much higher supraphysiologic CCK blood concentration is required for hormonal effect of CCK on the acinar cells. Intestinal serotonin also similarly triggers a vago-vagal reflex that augments and synergizes CCK action.

Recently the regulation of CCK secretion was found to involve two additional small peptide players; a duodenal derived CCK-releasing peptide (CCK-RP) and a 'monitor peptide' from the acinar cells. Fatty acids and amino acids in the chyme release CCK-RP that then stimulates CCK secretion.

The release of acinar 'monitor peptide' is also neurally mediated by vagal cholinergic and GRPnergic fibers before food enters the stomach. The vago-vagal reflexes during the gastric and intestinal integrated response to a meal also secrete monitor peptide. The CCK that is released by amino acids and fatty acids participates in the choreography of vago-vagal reflex of pancreatic juice secretion that includes the monitor peptide.

The two small peptides, remain undegraded and active when present among the large quantities of protein products in fresh gastric chyme that are digested by pancreatic proteases. When the dietary protein digestion is completed, the two regulatory peptides are themselves broken down and inactivated by the proteases. The CCK release is then terminated. Thus the neurally-stimulated monitor peptide provides a signal that helps to match the CCK-triggered pancreatic secretion to the intestinal digestive status.

6 What Is Unique About the Neural Control of Salivation?

Answer The salivary glands appear to be only regulated by neural inputs and are little affected by gastrointestinal hormones. In addition, both arms of the autonomic nervous system stimulate salivary flow, with a predominant effect by the parasympathetic fibers.

Concept The salivary glands, like the pancreas are made up of grape-like clusters of acinar cells that drain their secretions into a system of ducts that eventually empty into the oral cavity. The acini and ducts are ensheathed by contractile myofibroblasts that help to provide a hydrostatic pressure to expel saliva during high rates of secretion.

The salivary glands can sustain a very high rate of blood flow during salivation, more than, on a weight basis, than that observed in exercising skeletal muscles. This vasodilation effect by parasympathetic stimulation is not sensitive to atropine, which blocks cholinergic muscarinic receptors. The non-cholinergic fibers release VIP. In addition, kallikrein is secreted from the acinar cells into the interstitial fluid. Kallikrein, an enzyme, acts on kininogen to produce a potent vasodilator, bradykinin.

Parasympathetic stimulation produces a watery large volume saliva that is rich in amylase and mucins. The parasympathetic fibers are conveyed in the facial (VII) and glossopharyngeal (IX) nerves. The cholinergic fibers release acetylcholine that binds to muscarinic receptors on the acinar and ductal cells. The intracellular reactions activated include the production of inositol triphosphate (IP_3) and increased cytosolic calcium.

Sensory receptors in the mouth, pharynx and olfactory region send afferent signals to the salivatory nuclei in the medulla of the brain stem. The medullary salivation center also receives both facilitatory and inhibitory impulses from the hypothalamic feeding/appetite center neurons and from the cerebral cortex. Conditioned reflex of salivation, described by Pavlov are triggered by the sight, sound, smell and even the memory of a dish of delicious sea (see!) food. Conversely, stress, fear, sleep diminishes salivation.

Sympathetic stimulation also increases acinar secretion. However blood flow to the salivary glands is reduced due to an alpha-adrenergic vasoconstrictor action. Thus a low volume saliva results and a dry mouth is experienced during any sympathetic 'fight or flight' situations. The sympathetic input to the salivary glands originates in the thoracic segments (T1-T3) with pre-ganglionic fibers that synapse in the superior cervical ganglion. Post-ganglionic sympathetic fibers secrete noradrenaline that binds to beta adrenergic receptors on the acinar and ductal cells. The intracellular signaling initiated by beta-adrenergic action is the generation of cyclic adenosine monophosphate (cAMP).

Nausea is accompanied by increased salivation at the beginning of a vomiting reflex. This may be seen as a protective response ahead of the entry of acidic gastric vomitus into the esophagus and oral cavity.

Ductal secretion of bicarbonate increases during parasympathetic activity to the glands. The pH of saliva rises from 6.0 to about 8.0 with increased salivary flow. Patients with diminished salivary secretions have a reduced oral pH with greater associated tooth decay and esophageal injury.

7 If You Are Asked to Swallow Your Saliva, Does the Swallowing Process Involve Both Voluntary and Involuntary Events?

Answer The journey of food or liquid from the oral cavity to entry into the stomach is physiologic swallowing (deglutination) and includes both voluntary actions of skeletal muscles and autonomic mechanisms that control smooth muscles.

Concept For solid foods, mastication and salivary lubrication form the food bolus that is swallowed. The oral phase of deglutination is voluntary but the pharyngeal and esophageal phases of swallowing are involuntary, autonomic reflex events (although 'unconscious, involuntary' swallowing of saliva occurs numerous time during a 1 h lecture!).

Neurons in the brainstem mediate the involuntary phase of swallowing. The pharynx is enriched with mechanoreceptors. Afferent sensory fibers travel in the glossopharygeal (IX cranial nerve) to the NTS (nucleus of the tractus solitarius). Reflex motor efferent impulses are transmitted in the IX nerve and X nerve (vagus) to the pharynx, soft palate, upper oesophagus.

In the vagus nerve, besides the motor fibers that serve the skeletal muscles of the pharynx and larynx, the rest are almost all parasympathetic fibers. The soft palate is raised to prevent food entering the nasopharynx. The upper esophageal sphincter is relaxed. The larynx also rises so that the trachea is closed off by the epiglottis. Momentary deglutination apnea also prevents food from entering the trachea.

A wave of peristalsis is initiated by the swallowing center neurons in the brainstem. The esophageal peristaltic wave proceeds for about 10 s to propel the food bolus into the stomach. Any food particles that are not moved by the first peristaltic contraction are evacuated by a secondary esophageal peristalsis. This follow-up esophageal smooth muscle activity is triggered by a vago-vagal reflex as well as activation of local, enteric neural system (ENS) reflexes.

The lower esophageal or gastro-esophageal sphincter (LES) prevents reflux from the acidic contents of the stomach. The intra-esophagal pressure is subatmospheric within the thoracic cavity and the LES guards and protects the esophagus from acid erosion ('heartburns'). Relaxation of the LES that allows food to be 'swallowed' into the stomach is mediated by peptidergic vagal fibers which neurotransmit VIP (vasoactive -intestinal peptide). Nitric oxide nerve fibers are also likely LES muscle relaxants.

The smooth muscle tone of the orad region of the stomach is at the same time also reduced by a neural vagally-produced receptive relaxation. This allows accommodation of swallowed food without a large increase in intra-gastric pressure.

Achalasia is a condition when the esophagus is greatly dilated due to failure of LES opening. There is a histological lack in the myenteric plexus of the esophageal ENS intrinsic neural network.

8 How Does Stress Likely Promote Gastric/Duodenal Ulcerogenesis?

Answer Chronic experience of stress probably tips the physiologic balance that protects against the injury and development of gastro-duodenal ulcers. This imbalance is between the rate of gastric juice secretion and the protective mechanisms at the gastro-duodenal mucosal surface.

Concept Peptic ulceration is prevented by the physical gastric mucosal barrier and the mechanisms of acid neutralization of gastric chyme in the duodenum.

The gastric mucosa is lined by a viscous mucus layer whose acid-resistant property is enhanced by a bicarbonate-rich thin fluid layer sandwiched between the mucus and the cell surface. Local paracrine prostaglandins are involved in the secretion and generation of this muco-alkaline layer.

In addition, the extremely adhering tight junctions linking gastric epithelial cells prevent diffusion of acid and gastric pepsin to come into contact with the cells. Increased acidity in the empty stomach stimulates the paracrine somatostatin which then turns off gastrin secretion from the antral endocrine cells.

In the duodenum, the physical cell barrier is less restrictive than in the stomach. The Brunner's glands of the upper duodenum secrete an alkaline mucus. The bile that is secreted in response to a meal, stimulated by hormones CCK and secretin also contains bicarbonate. Secretion of pancreatic juice enriched with bicarbonate is stimulated by secretin, released from the duodenal endocrine cells by the acidic chyme.

The acidic chyme in the duodenum also triggers local and perhaps long neural vagal reflexes that slow down gastric emptying. There is thus a controlled balance between the needed secretion of gastric acid-pepsin juice for digestion and the self-(cell-)protecting physiologic properties and pathways at the gastro-duodenal mucosa.

Regular stresses involve activation of the hypothalamic-pituitary-adrenocortical axis. The major stress-adaptive glucocorticoid, cortisol has effects in reducing the gastro-protective mucosal barrier. This could shift the balance and cause pro-ulcerogenic reactions. The use of non-steroidal anti-inflammatory drugs (NSAID) was designed to diminish this pharmacological contraindication. Drinking alcohol is used to numb and relieve psychogenic stress. Unfortunately, alcohol can weaken the mucosal barrier properties.

The autonomic neural activities during chronic stresses could be distinguished from the classic 'fight or flight' responses which are presentations of increased sympathetic general discharge. Gastric acid secretion is increased indirectly by vagal parasympathetic activity to gastrin-, histamine-secreting and directly on HCl producing parietal cells. Pepsin secretion is also increased by parasympathetic action. Presumably, increased stress can eventually cause peptic ulceration as a result of inappropriate, excess secretion of gastric acid-pepsin juices by overactive parasympathetic activity to the stomach. Vagotomy is however no longer a recommended, surgical approach in peptic ulcer treatment.

The most frequent ulcerative sites are within a few centimeters from the pylorus. Peptic ulcers also more frequently develop along the lesser curvature of the antral end of the stomach.

9 How Is the Vagus Nerve Involve in the Digestion of Lipids in Our Diet?

Answer The vagus nerve contracts the gall bladder and relaxes the sphincter of Oddi to propel bile in to the duodenum for the emulsification of lipids in the gastric chyme. The vagal parasympathetic nerve also stimulates pancreatic juice that contains lipase that digests the lipids.

Concept The action of the vagal parasympathetic nerve is increased during eating and digestion. The vagal activity is activated by central nervous neurons as well as by reflexes triggered from activity in the gastrointestinal tract.

The hormone cholecystokinin (CCK) is released from the duodenum when lipid and protein products in chyme enters the duodenum. CCK causes contraction of the gall baldder hormonally. Interestingly CCK also stimulates vagal afferent sensory fibers in the duodenal wall. This results in a vago-vagal, long loop reflex that enhances the gall baldder contraction via cholinergic vagal efferents. Post-prandial gall bladder contraction is co-ordinated and timed to coincide with gastric emptying. The hormone CCK also controls (reduces) gastric emptying to allow for optimal duodenal emulsification and digestion of lipids.

The vago-vagal reflex simultaneously also leads to relaxation of the spincter of Oddi to allow bile flow into the intestine. This CCK-induced vagal reflex releases enteric neural transmitters like VIP and nitric oxide that inhibit the smooth muscle tone of the Oddi sphincter.

The pancreatic exocrine secretion is also stimulated by parasympathetic vagal fibers. Pancreatic lipase acts on triglycerides to produce fatty acids and monoglycerides. The pancreatic acinar cells also co-secrete a co-lipase which is needed to stabilize the lipase on the lipid oil droplets for continual lipolytic digestion. Bicarbonate-rich pancreatic juice is also needed to neutralize the gastric chyme since the pancreatic lipase has a pH optimum around 7.0.

The liver is responsiBILE for the production and secretion of bile acids. Bile acids are amphipathic metabolic products of cholesterol. Bile acids are actively secreted across the canalicular membrane of the hepatocyte. Phosphatidylcholine, a component of the hepatocyte cell membrane is selectively released into the lumen and form mixed micelles with the bile acids. Cholesterol and conjugated bilirubin also actively enter into the canalicular bile.

The enterohepatic cycling of bile acids regulates the rate at which they are synthesized and transported. Bile acids have a feedback inhibition on the initial, rate-limiting step of 7-hydroxylation of cholesterol. When bile acids are recycled from the intestinal ileum, new synthesis of bile acids is decreased. If however, the entero-hepatic circuit is interrupted, more cholesterol is freshly converted to bile acids. The liver is the only organ solely responsiBILE for the excretion or removal of cholesterol from the body, whether as bile acids or as intact, native cholesterol in the bile flow.

Thus, we could say that the vagus parasympathetic nerve is involved in cholesterol homeostasis, by its action on the gall bladder/sphincter of Oddi, directly or vago-vagally activated by CCK.

10 What Are the Sensory Afferents Trigger the Vomiting Reflex?

Answer Afferent inputs from the gastrointestinal tract, the vestibular apparatus and higher cortical centers can all induce the vomiting responses.

Concept Vomiting highlights a central control of gastrointestinal motility. It obviously involves a reverse peristalsis and upper intestinal contents are moved into the stomach. It is preceded commonly with anorexia, begins with a sensation of nausea and increased salivation. The glottis closes over the trachea to prevent aspiration of vomitus into the airways.

The intra-abdominal muscles (skeletal) are contracted, raising the intra-abdominal pressure. The lower esophageal sphincter as well as the esophageal smooth muscles are relaxed and the gastric contents are then strongly ejected. Retching refers to the physical events leading to vomiting just before the intra-thoracic pressure becomes great enough to force the vomitus through the upper esophageal sphincter.

These co-ordinated multitasks of the vomiting reflex, effected through smooth and skeletal muscle activities are orchestrated by neurons in the reticular formation of the medulla in the brain stem. Mucosal irritation or abnormal luminal distention of the GI tract (e.g. intestinal blockage due to a malignancy) is a peripheral trigger. Afferent signals are conveyed via vagal parasympathetic and sympathetic nerves. The gag reflex is produced by pharyngeal stimulation relayed via the glosso-pharygeal nerve.

Central activation of vomiting can arise as in the common association of nausea and vomiting in motion sickness due to disorientated vestibular afferent inputs to the vomiting center. Pro-vomiting or emetic responses are also experienced by some persons, caused by unpleasant smells or uncomfortable sights. Elevated intra-cranial pressure resulting e.g. from a cerebral hemorrhage also causes vomiting.

Medullary chemoreceptor sensing is another pathway that initiates vomiting when stimulated by circulating agents (the central chemoreceptors that control respiration is another group of medullary cells). This pro-emesis chemical-sensitive site is called ‘chemoreceptor trigger zone’ and is located in the area postrema, a structure of the circumventricular organs that is not guarded by the blood brain barrier. The side effect of drugs that produce emesis acts in this chemo-trigger zone. The symptoms of nausea/vomiting in pregnancy, uremia of kidney failure and radiation sickness are also mediated by these area postremal cells.

There are dopamine and serotonin (5-HT)receptors in the area postrema and the nucleus tractus solitarius that when activated induce vomiting. Dopamine and 5-HT antagonists are prescribed as emetics.

The vomiting reflex can in some aspects be viewed as an adaptive response to rid the body of potentially toxic chemicals ingested in foods. Vomiting dehydrates the person, the loss of gastric acid resulting in a metabolic alkalosis made worse by a ‘volume contraction alkalosis’. Hypokalemia is also present.

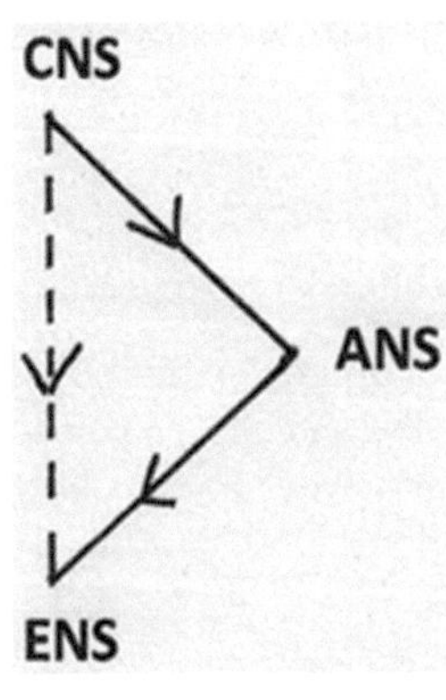

The central nervous system (CNS) is linked to the enteric nervous system (ENS) of the GI tract via the parasympathetic and sympathetic divisions of the autonomic nervous system (ANS)

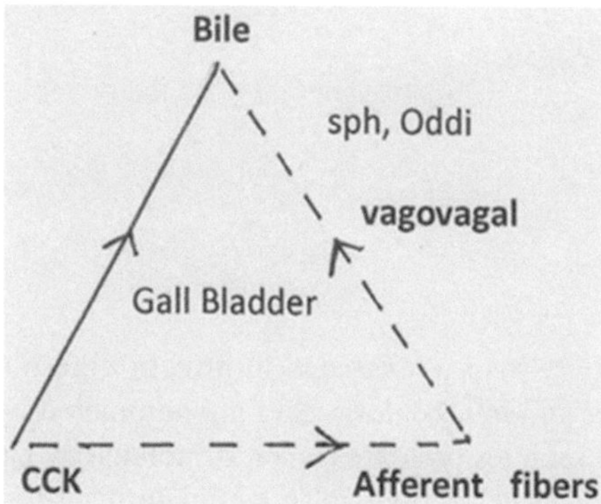

Fatty chyme from the stomach stimulates secretion of the hormone CCK from the duodenum. The CCK contracts the gall bladder and relaxes the sphincter of Oddi to release bile flow into the duodenum. CCK also has a neuro-endocrine effect in stimulating afferent vagal fibers. Then vago-vagal reflexes augment the expulsion of gall bladder bile for lipid digestion and absorption in the small intestines.

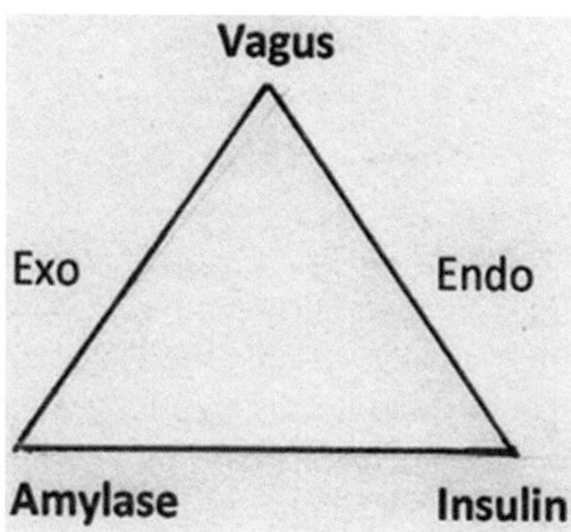

The parasympathetic vagus nerve supplies both the pancreatic acinar cells and the endocrine beta cells. Vagal input to glucose homeostasis thus includes the secretion of pancreatic amylase for carbohydrate digestion and the secretion of insulin during cephalic and gastric phases of a meal.

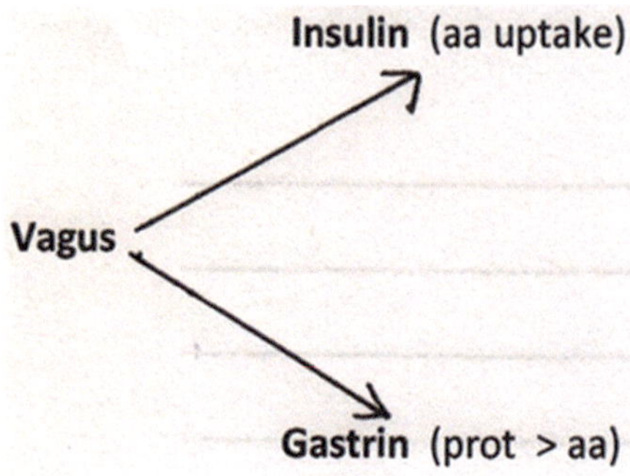

The vagal nerve activity also participates in protein digestion and usage. In the stomach, both gastrin and hydrochloric acid are stimulated by vagal activity. The pancreatic acinar cell secretes proteases when stimulated by parasympathetic nerve action. Cholinergic vagal nerve fibers also stimulate insulin release. Amino acids stimulate secretion of insulin and this anabolic hormone then increases the cellular uptake of amino acids

Using playing cards to think through integration in Physiology. The Hearts would naturally be used for cardiovascular system. The Clubs, turned upside down, can visually represent the respiratory tree and the alveoli. The Diamonds which resemble drops of liquid and urine (not bloody !) would tie in with the Renal system. The Spades would mix and shovel, moving things…can remind us of the Gastrointestinal system (the GI and renal functions share many common epithelial absorptive and secretory membrane mechanisms). *The Neuro-endocrine systems would not be left out of our Physio card games. Well, the Spade looks like 2 connecting neurons! And the Clubs could represent the secretory glands and cells that secrete the hormones!*

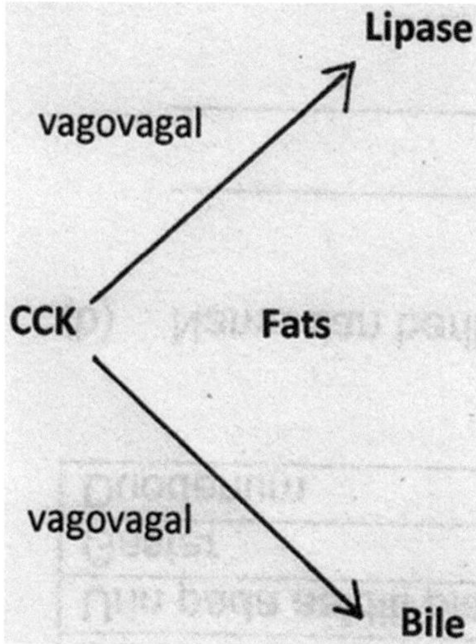

Cholecystokinin (CCK) has been shown to stimulate afferent fibers of the vagus nerve. This then triggers a vago-vagal reflex that contracts the gall bladder besides the direct action of CCK on the smooth muscle of the gall bladder. This local activation of vagal afferents by CCK also leads to a vago-vagal reflex stimulation of pancreatic juice that contains amylase, lipolytic and proteolytic enzymes.

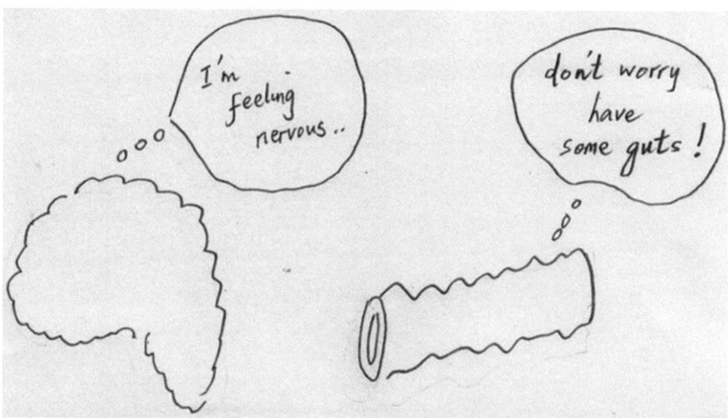

The central nervous system and the enteric neural network in the GI tract are interconnected. We can think of 'nervous diarrhoea' and sensation of satiety conveyed by afferent signals from mechano-receptors in the GI wall.

Appendix

These selected eponyms were compiled by Dr Stefanie Cheang, a graduate of Perdana Univ -John Hopkins Medical Programme, Malaysia. Stefanie was my biomedical student when she completed her essay assignment on the physiology of denervated kidneys. An appreciation of the historical heritage of current knowledge will enhance students' learning of physiology. This could be called 'Phystory'.

H.M. Cheng (ed.), *Physiology Question-Based Learning*,
DOI 10.1007/978-981-10-0877-1

Name		Nationality & occupation	University attended	Eponym associated with	First reference	Year
Adamkiewicz, Albert Wojciech	(1850–1921)	Polish pathologist	Jagiellonian University	Described the blood supply of the human spinal cord (an anterior radicular artery supplying the lumbar region of the spinal cord known as the artery of Adamkiewicz)	"Die Blutgefasse des menschlichen Ruckenmarkes-die gefasse der Ruckenmarksubstanz"	1881
					"Die Gefasse der ruckenmarksoberflache" in 1882	1882
Adie, William John	(1886–1935)	English neurologist	University of Edinburgh	Described Adie's Syndrome which is a clinical syndrome involving a tonic pupil with diminished or absent deep tendon reflexes	Adie WJ. Tonic pupils and absent tendon reflexes: a benign disorder sui generis; its complete and incomplete forms. Brain 1932; 55: 98–113	1932
Alzheimer, Alois	(1884–1915)	German neuropsychiatrist	Universities of Berlin, Tübingen, and Würzberg	Studied presenile and senile dementia, and described the condition now known as Alzheimer's disease	Alzheimer A. Über einen eigenartigen schweren Erkrankungsprozeß der Hirnrinde. Neurologisches Centralblatt 1906; 23: 1129–36.	1906
Argyll Robertson, Douglas Moray Cooper Lamb	(1837–1909)	Scottish ophthalmologist	University of St Andrews	The Argyll Robertson pupil includes, among other signs, pupillary constriction in accommodation, but not in response to light	Robertson DA (1869) On an interesting series of eye symptoms in a case of spinal disease, with remarks on the action of belladonna on the iris. Edinb Med J 14:696–708 Robertson DA (1869) Four cases of spinal myosis with remarks on the action of light on the pupil. Edinb Med J 15:487–493	1869

Auerbach, Leopold	(1828–1897)	German anatomist	University of Breslau	Discovered the Auerbach's plexus, a layer of ganglion cells that provide control of movements of the gastro-intestinal tract, also known as the "myenteric plexus"	Auerbach L. Vorläufige Mitteilung. Breslau: E. Morgenstern; 1862. Über einen Plexus myentericus, einen bisher unbekannten ganglio-nervösen Apparat im Darmkanal der Wirbeltiere.	1862
Babinski, Joseph François Félix	(1857–1932)	French clinical neurologist of Polish origin	University of Paris	Described the Babinski sign, which consists of up-turning of the great toe and spreading of the toes on stroking the sole, is characteristic of an upper motor neuron lesion	Babinski,]. (1896 a). Sur le réflexe cutané plantaire dans certaines affections organiques du système nerveux central. Comptes rendus des Séances de la Société deBiologie,48,207-208	1896
Bainbridge, Francis Arthur	(1874–1921)	British physiologist	Trinity College, Cambridge	Discovered that increase of pressure on the venous side of the heart accelerates the heart rate, now known as the Bainbridge Reflex	Bainbridge FA. The influence of venous filling upon the rate of the heart. J Physiol 50: 65–84, 1915	1915
Bálint, Rezső	(1874–1929)	Hungarian clinical neurologist and psychiatrist	University of Budapest	Balint syndrome is a combination of visual disorientation, ocular apraxia and optic ataxia, due to bilateral destructive lesions in the superior parts of the occipital and parietal lobes	Bálint, R (1909). "Seelenlähmung des 'Schauens', optische Ataxie, räumliche Störung der Aufmerksamkeit" (pdf). European Neurology 25 (1). pp. 51–66, 67–81	1909

(continued)

Name		Nationality & occupation	University attended	Eponym associated with	First reference	Year
Bell, Sir Charles	(1774–1842)	Scottish anatomist, clinical neurologist, and surgeon.	University of Edinburgh	The Bell-Magendie law states that dorsal spinal roots are sensory, whereas ventral roots are motor	Magendie F. Note sur le siège du mouvement et du sentiment dans la moelle épinière. J Physiol expér Path. 1822;2:366–371.	1822
Betz, Vladimir A.	(1834–1894)	Russian anatomist	St Vladimir University in Kiev	Discovered and described the giant pyramidal cells (Betz cells) in the motor area of the cerebral cortex	Betz W. (1874) Anatomischer Nachweis zweier Gehirncentra. Centralblatt für die medizinischen Wissenschaften. 12:578–580, 595–599.	1874
Breuer, Josef	(1842–1925)	Austrian physician and psychologist	University of Vienna	Demonstrated the role of the vagus nerve in the reflex nature of respiration, a mechanism now known as the Hering–Breuer reflex	Hering, E., and Breuer, J., Sitzungsber. k. Akad. Wissench., Math.- naturw. Cl., Wien, 1868, lvii, pt. 2, 672; Iviii, pt. 2, 909.	1868
Broca, Pierre Paul	(1824–1880)	French pathologist and anthropologist		Broca's Area is the area of the left hemisphere of the brain at the posterior end of the Gyrus frontalis inferior (area 44 and 45) of the dominant hemisphere. Broca's aphasia (also known as Motor aphasia or Expressive aphasia), which is characterized by halted, fragmented, effortful speech, but relatively well-preserved comprehension	Broca MP. Remarques sur le siége de la faculté du langage articulé, suivies d'une observation d'aphemie (Perte de la Parole) Bulletins et Memoires de la Societe Anatomique de Paris. 1861;36:330–357	1861
Brodmann, Korbinian	(1868–1918)	German neuropsychiatrist	University of Berlin	Brodmann area is a numbered and mapped out region of the cerebral cortex, in the brain, defined by its cytoarchitecture, or histological structure and organization of cells	Brodmann K (1909). "Vergleichende Lokalisationslehre der Grosshirnrinde" (in German). Leipzig: Johann Ambrosius Barth	1909

Brown-Séquard, Charles Edouard	(1817–1894)	Mauritan physiologist and clinical neurologist	Charité	Brown-Séquard syndrome consists of the sensory and motor abnormalities that follow hemisection of the spinal cord	C.-É. Brown-Séquard: De la transmission croisée des impressions sensitives par la moelle épinière. Comptes rendus de la Société de biologie, (1850)1851, 2: 33-44	1850
Bucy, Paul Clancy	(1904–1992)	American neurosurgeon	University of Iowa University of Chicago Medical School University of Illinois Medical School Northwestern University Medical School	Klüver–Bucy syndrome is a syndrome resulting from bilateral lesions of the anterior temporal lobe (including amygdaloid nucleus)	Klüver, H.; Bucy, P. C (1937) "Psychic blindness" and other symptoms following bilateral temporal lobectomy in Rhesus monkeys. American Journal of Physiology, Vol 119, 352–353.	1937
Ramón y Cajal, Santiago Felipe	(1852–1934)	Spanish histologist	University of Zaragoza	Cajal vigorously championed the Neuron Doctrine on the basis of his observations with silver staining methods Interstitial cell of Cajal (ICC).	Cajal SR. Histologie du système nerveux de l'homme et des vertébrés. Vol. 2. Paris: Maloine; 1911. pp. 891–942	1911

(continued)

Name		Nationality & occupation	University attended	Eponym associated with	First reference	Year
Chiari, Hans	(1851–1916)	Austrian pathologist	University of Prague	Arnold-Chiari malformation: A condition in which the inferior poles of the cerebellar hemispheres (cork-like protrusions) and the medulla oblongata protrude through the foramen magnum into the spinal canal, without displacing the lower brain stem	Über Veränderungen des Kleinhirns infolge von Hydrocephalie des Grosshirns Deutsche medicinische Wochenschrift, Berlin, 1891, 17: 1172–1175	1891
					Über Veränderungen des Kleinhirns, der Pons und der Medulla oblongata, infolge von congenitaler Hydrocephalie des Grosshirns Denkschriften der Akademie der Wissenschaften in Wien, 1895, 63: 71	1895
Clarke, Jacob Augustus Lockhard	(1817–1880)	English anatomist and clinical neurologist	Guy's Hospital St Thomas' Hospital.	Described the nucleus dorsalis (thoracicus) of the spinal cord, known as Clarke's column	J. A. L. Clarke: Researches into the structure of the spinal cord Philosophical Transactions of the Royal Society of London, 1851, 141: 607–621	1851
Cushing, Harvey	(1869–1939)	American neurosurgeon	Yale University Harvard Medical School	Cushing contributed to many basic aspects of neurology, including the function of the pituitary gland, pituitary tumors, tumors of the eighth cranial nerve, and classification of brain tumors Cushing reflex: a physiological nervous system response to increased intracranial pressure (ICP) that results in Cushing's triad of increased blood pressure, irregular breathing, and a reduction of the heart rate	Cushing, H (1901). "Concerning a definite regulatory mechanism of the vasomotor centre which controls blood pressure during cerebral compression". Bull Johns Hopkins Hosp. 12: 290–2	1851

Golgi, Camillo	(1843–1926)	Italian histologist	University of Pavia	Introduced a silver staining method that provided the basis of numerous advances in neurohistology. Described type I and type II neurons and the tendon spindles, and the organelle now called the Golgi apparatus Awarded the Nobel Prize for Medicine and Physiology in 1906 (with Ramón y Cajal) Golgi tendon organ: A proprioceptive sensory receptor organ that is at the origins and insertion of skeletal muscle fibers into the tendons of skeletal muscle. It provides the sensory component of the Golgi tendon reflex	Golgi, C. (1903) Opera Omnia, Vol. I-III. R. Fusari, G. Marenghi G. and L. Sala (eds), Hoepli Editore, Milano (Describes the Golgi Tendon Organ)	1878
Hirschsprung, Harald	(1830–1916)	Danish physician	Queen Louisa Hospital for Children	Described Hirschprung's Disease which is a condition with a megacolon due to lack of ganglion cells with failure of development of the myenteric plexus of the rectosigmoid area of the large intestine	Hirschsprung, H. (1888). "Stuhlträgheit Neugeborener in Folge von Dilatation und Hypertrophie des Colons". Jahrbuch für Kinderheilkunde und physische Erziehung (Berlin) 27: 1–7	1888
Huntington, George Sumner	(1850–1916)	American general medical practitioner	University of Columbia, New York	Huntington described a hereditary form of chorea resulting from neuronal degeneration in the corpus striatum and the cerebral cortex	G. Huntington: On Chorea. Medical and Surgical Reporter of Philadelphia, volume 26, no. 15, April 13, 1872, pp. 317–321	1872
Lissauer, Heinrich	(1861–1891)	German neurologist	Universities of Heidelberg, Berlin and Leipzig	Described the dorsolateral fasciculus of the spinal cord (Lissauer's tract or zone)	Beitrag zur pathologische Anatomie des Tabes dorsalis und zum Faserverlauf in menschlichen Rückenmark. Neurologisches Centralblatt, 1885, 4: 245–246	1885
Luschka, Hubert von	(1820–1875)	German anatomist	University of Tübingen	Luschka described the lateral apertures of the fourth ventricle (foramina of Luschka; also called the foramina of Key and Retzius)	von Luschka H (1855). Die Adergefletche der Menschlichen Gehirnes. Eine Monographie. Berlin: Reimer	1855

(continued)

Name		Nationality & occupation	University attended	Eponym associated with	First reference	Year
Magendie, François	(1783–1855)	French physiologist	College of France	The sensory function of dorsal spinal nerve roots and motor function of ventral roots constitute the Bell-Magendie law Magendie also described the median aperture of the fourth ventricle (foramen of Magendie)	Magendie, Francois (1822). "Expériences sur les fonctions des racines des nerfs rachidiens". Journal de physiologie expérimentale et de pathologie (in French): 276–9 Magendie F (1828). Memoire Physiologique sur le Cerveau. Journal de Physiologie Experimentale et de Pathologie 8: 211–218	1822 1828
Meissner, Georg	(1829–1905)	German anatomist and physiologist	University of Göttingen	His name is associated with Meissner's corpuscles, which are tactile corpuscles of the skin. His name is also associated with Meissner's plexus which are the submucosal nerve plexus of the gastrointestinal tract	R. Wagner and G. Meissner: Ueber das Vorhandensein bisher unbekannter eigenthümlicher Tastkörperchen (Corpuscula tactus) in den Gefühlswärzchen der menslichen Haut und über die Endausbreitung sensitiver Nerven Nachrichten von der Georg-Augusts-Universität und der Königlichen Gesellschaft der Wissenschaften zu Göttingen, Heft No. 2 of February 2, 1852: 17–30. Göttingen, 1852 G. Meissner: Über die Nerven der Darmwand. Zeitschrift für rationelle Medizin, n.s. 1857, 8: 364–366	1852 1857

Ménière, Prosper	(1801–1862)	French otologist	Hôtel-Dieu de Paris	Described Menière's disease; the syndrome characterized by episodes of vertigo, nausea, and vomiting occurring in some diseases of the internal ear	Méniere (1861) "Sur une forme de surdité grave dépendant d'une lésion de l'oreille interne" (On a form of severe deafness dependent on a lesion of the inner ear), Bulletin de l'Académie impériale de médecine, 26: 241	1861
Merkel, Friedrich Siegmund	(1845–1919)	German anatomist	University of Erlangen	Described tactile endings in the epidermis, that were subsequently given the eponym "Merkel cells" in 1878 by Robert Bonnet	Merkel FS. (1875). Tastzellen und Tastkörperchen bei den Hausthieren und beim Menschen. Archiv für mikroskopische Anatomie, 11: 636–652	1875
Meyer, Adolph	(1866–1950)	American psychiatrist	University of Zurich Johns Hopkins University	The fibers of the geniculocalcarine tract that loop forward in the temporal lobe constitute Meyer's loop	Meyer, A. (1907). The connections of the occipital lobes and the present status of the cerebral visual affections. Trans Ass Am Phys,22, 7–16	1907
Monro, Alexander	(1733–1817)	Scottish anatomist	University of Edinburgh	Known for the Monro-Kellie doctrine on intracranial pressure, a hypothesis developed by Monro and his former pupil George Kellie, who worked as a surgeon in the port of Leith Described the communication between the lateral and third ventricles of the brain known as 'foramen of Monro' and described the changes seen in hydrocephalus	Monro A (1783). Observations on the structure and function of the nervous system. Edinburgh: Creech & Johnson	1783

(continued)

Name		Nationality & occupation	University attended	Eponym associated with	First reference	Year
Müller, Johannes Peter	(1801–1858)	German physiologist	University of Bonn	Proposed the **Muller's doctrine of specific nerve energies** which holds that esch sensory nerve, however stimulated, gives rise to only one type of sensory process and a single quality of sensation	Muller J. (1833–1840) Handbook des physiologie de menschen. 3 vols. Coblenz: Holscher	1826
Nissl, Franz	(1860–1919)	German neuropsy-chiatrist	Ludwig Maximilian University of Munich	Nissl introduced a method of staining gray matter with cationic dyes to show the basophil material (Nissl bodies) of nerve cells	F. Nissl: Über die sogenannten Granula der Nervenzellen Neurologisches Centralblatt, Leipzig, 1894, 13: 676–685, 781–789, 810–814	1894
Parinaud, Henri	(1844–1905)	French ophthal-mologist.	Paris	Described Parinaud syndrome which is a syndrome of paralysis of upward gaze (with sparing of convergence) is due to a lesion of the midbrain, typically due to pressure from a pineal tumor	Parinaud, H. Paralysis of the movement of convergence of the eyes. Brain. 1886;9:330–341	1886
Parkinson, James	(1775–1824)	English physician, surgeon, and paleontologist	Barts and The London School of Medicine and Dentistry	Parkinson described "shaking palsy" or paralysis agitans. Jean-Martin Charcot coined the term "Parkinson's disease" some 60 years later	An Essay on the Shaking Palsy	1817
Purkinje, Johannes (Jan) Evangelista	(1787–1869)	Bohemian physiologist, pioneer in histological techniques, and an accomplished histologist	Charles University in Prague University of Breslau in Prussia	Described the Purkinje cells of the cerebellar cortex	Purkinje, J., Versammlung der Naturforscher und Aertze zu Prag im September 1837, Isis yon Oken, 1838, 7,581.	1837

Ranvier, Louis-Antoine	(1835–1922)	French histologist and a founder of experimental histology	Collège de France	Ranvier described the nodes of Ranvier in the myelin sheaths of nerve fibers	L. A. Ranvier: Leçons sur l'histologie du système nerveux. 2 volumes. Paris, F. Savy, 1878	1878
Romberg, Moritz Heinrich	(1795–1873)	German clinical neurologist		Romberg's sign of impaired proprioceptive conduction in the spinal cord consists of abnormal unsteadiness when standing with the feet together and the eyes closed	M. H. von Romberg: Tabes dorsalis. In his Lehrbuch der Nervenkrankheiten des Menschen Berlin, 1846; volume I, page 795	1846
Sherrington, Sir Charles Scott	(1856–1952)	English neurophy-siologist	Royal College of Surgeons of England	**Liddell-Sherrington reflex** Tonic contraction of the muscles in response to their being stretched **Schiff-Sherrington reflex** A grave sign in animals: rigid extension of the forelimbs after damage to the spine **Sherrington's law I** Every posterior spinal nerve root supplies a particular area of the skin **Sherrington's law II** The law of reciprocal innervation **Vulpian-Heidenhain-Sherrington phenomenon** Slow contraction of denervated skeletal muscle by stimulating the autonomic cholinergic fibres innervating its blood vessels	Sherrington CS. The integrative action of the nervous system. New Haven, CT: Yale University Press; 1906	1906
Schwann, Theodor	(1810–1882)	German anatomist	Jesuits College in Cologne University of Bonn	Formulated the Cell Theory (with M. J. Schleiden) and described the neurolemmal cells (Schwann cells) of peripheral nerve fibers	T. Schwann: Mikroskopische Untersuchungen über die Übereinstimmung in der Struktur und dem Wachstum der Thiere und Pflanzen. Berlin, Sander, 1839	1839

(continued)

Name		Nationality & occupation	University attended	Eponym associated with	First reference	Year
Wallenberg, Adolf	(1862–1949)	German physician	Leipzig University	Described the lateral medullary syndrome (Wallenberg's syndrome)	Wallenberg A (1901) Anatomischer Befund in einen als acute Bulbäraffection (Embolie der Art. cerebellar post. sinistr) beschriebenen Falle. Arch Psych Nervenkrankh 34:923–959	1901
Waller, Augustus Volney	(1816–1870)	English physician and physiologist	Queen's College, Birmingham	Described the degenerative changes in the distal portion of a sectioned peripheral nerve, known as wallerian degeneration	Waller, A. (1850). "Experiments on the Section of the Glossopharyngeal and Hypoglossal Nerves of the Frog, and Observations of the Alterations Produced Thereby in the Structure of Their Primitive Fibres". Philosophical Transactions of the Royal Society of London 140 (0): 423–429	1850
Weber, Sir Hermann David	(1823–1918)	English physician	Marburg Medical School	Described the midbrain lesion causing hemiparesis and ocular paralysis. The term Weber's syndrome was introduced by Joseph Grasset	J. Grasset: Un type spécial de paralysie alterne motrice etc. Revue neurologique, Paris, 1900, 8: 586	1900

Wernicke, Carl	(1848–1905)	German neuropsy-chiatrist	Charité, University of Breslau, University of Halle	Wernicke's sensory language area and Wernicke's aphasia (which consists of loss of comprehension of spoken language, loss of ability to read (silently) and write, and distortion of articulate speech) are named after him	K. Wernicke: Der aphasische Symptomencomplex. Eine psychologische Studie auf anatomischer Basis Breslau, M. Crohn und Weigert, 1874 *(describes Wernicke's Aphasia)*	1874
Westphal, Karl Friedrich Otto	(1833–1890)	German clinical neurologist	Charité	Westphal described the Edinger-Westphal nucleus in the oculomotor complex	C. F. O. Westphal: Ueber einen Fall von chronischer progressiver Lähmung der Augenmuskeln (Ophthalmoplegia externa) nebst Beschreibung von Ganglienzellengruppen im Bereiche des Oculomotoriuskerns Archiv für Psychiatrie und Nervenkrankheiten, Berlin, 1887, 18: 846–871	1887
Wilson, Samuel Alexander Kinnier	(1878–1937)	British clinical neurologist	University of Edinburgh Medical School King's College Hospital	Described hepatolenticular degeneration, known as Wilson's disease	Kinnier Wilson SA (1912). "Progressive lenticular degeneration: a familial nervous disease associated with cirrhosis of the liver". Brain 34 (1): 295–507. doi:10.1093/brain/34.4.295	1912

Bibliography

Barrett, KE (2014) Gastrointestinal physiology, 2nd edn. New York: McGraw-Hill

Barrett KE, Barman SM, Boitano S, Brooks HL (2012) Ganong's review of medical physiology, 24th edn. New York: McGraw-Hill

Cheng HM (2013) Thinking through physiology. Kuala Lumpur Pearson

Hall J (2011) Guyton and hall textbook of medical physiology, 12th edn. Edinburgh: Elsevier

Michael-Titus A, Revest P, Shortland P (2010) The nervous system: basic science and clinical conditions, 2nd edn. Edinburgh: Elsevier

Raff H, Levitzky M (2011) Medical physiology: a systems approach. New York: McGraw-Hill

Sherwood L (2010) Human physiology: from cells to systems, 7th edn. Singapore: Cengage Learning

Widmaier EP, Raff H, Strang KT (2011) Vander's human physiology, 12th edn. New York: McGraw-Hill

H.M. Cheng (ed.), *Physiology Question-Based Learning*,
DOI 10.1007/978-981-10-0877-1

MIX
Papier aus verantwortungsvollen Quellen
Paper from responsible sources
FSC® C105338

If you have any concerns about our products,
you can contact us on
ProductSafety@springernature.com

In case Publisher is established outside the EU,
the EU authorized representative is:
Springer Nature Customer Service Center GmbH
Europaplatz 3, 69115 Heidelberg, Germany

Printed by Libri Plureos GmbH
in Hamburg, Germany